The Marine Fish Families Series

Seahorses . Pipefishes

and their relatives

A Comprehensive Guide to
Syngnathiformes

Rudie H Kuiter

TMC
publishing

TMC Publishing, Chorleywood, UK

TABLE OF CONTENTS

Kuiter, Rudie H: Seahorses, Pipefishes and their Relatives, a comprehensive guide to Syngnathiformes

First English language edition 2000. Revised 2003

Published by TMC Publishing, Chorleywood, UK

Design, editing and taxonomic decisions: Rudie H. Kuiter.

Print and production: Graspo CZ, a.s., Zlin.

TMC Publishing
Solesbridge Lane,
Chorleywood, Herts WD3 5SX
United Kingdom

Tel: +44 (0) 1923 284151 Fax: +44 (0) 1923 285840
Email: info@tmc-publishing.com
Website: www.tmc-publishing.com

ISBN 0-9539097-4-3

| *REEF* | *CORAL* | *WEED* | *SAND* |

| *SOLENOSTOMIDAE* 205 | *CENTRISCIDAE* 214 | *MACRORAMPHOSIDAE* 218 | *PEGASIDAE* 225 |

| *AULOSTOMIDAE* 222 | *GASTEROSTEIDAE* 229 |

| *FISTULARIIDAE* 223 | *AULORHYNCHIDAE* 230 |

Acknowledgements: I was privileged to know the late Chuck Dawson, the pipefish-guru, who taught me much about pipefish-taxonomy, and his work formed the platform for my book. I thank Martin Gomon, Museum Victoria, for his help and further advice on taxonomy.

I thank Mark Chettle, Zoonetics, for setting up the entire computer system and help with all the publication aspects.

Toshikazu Kozawa & Hiroyuki Tanaka provided photographs of many photographers in Japan.

My thanks to the contributing photographers. I'm particularly grateful for the behaviour photographs by Werner Fiedler, Tomonori Hirata and Patrick Louisy, but others provided important material that completed this works and each are individually acknowledged in the captions to the photographs.

Special thanks to Stuart Poss & MW. Littman, Gulf Coast Research Laboratory Museum, for making the Dawson material available.

INTRODUCTION

Seahorses are among the most unusual of all fishes and probably the most 'unfish'-like. They have an upright posture, a head bend at angle, a tubed snout with a small mouth at the tip, and a strong prehensile tail to grab onto things. A look that is far from the familiar scaly creature caught on line and hook. It is not surprising that many people find it difficult to believe that they are actually a kind of fish, one that belongs to the same phylum as the goldfish or herring, that at some stage were thought to be insects. However, seahorses are only the better-known members of a large diverse group, the pipefish family. These fishes have adapted to specific niches in reef or algal habitats, and a long time of evolution has created this numerous and highly diverse family. Besides the seahorses, this family encompasses seadragons, pipefishes and pipehorses, amongst which are some of the most fascinating and interesting examples of adaptations to nature.

The family is scientifically termed SYNGNATHIDAE, meaning 'jaw-fused', and its members are generally known as the PIPEFISHES because of their long tubular bodies. There are 55 genera that include more than 320 species, which vary in different postures, body shapes and fin arrangements. Based on shared features, the family has several identifiable groups that are here placed into four subfamilies, but some are pending further studies and presently included in SYNGNATHINAE for convenience. The latter encompasses the largest group, the actual pipefishes, which are mostly stick-like with their head in line with the body. A small caudal fin is usually present and males incubate eggs in pouch that is formed by simple or overlapping membranes under the trunk or tail. The HIPPOCAMPINAE comprises the seahorses and the pygmy pipehorses, which have a fully enclosed pouch with a small opening, for incubation of eggs, and prehensile tail. The SOLEGNATHINAE are the seadragons and pipehorses, in which the tail is more or less prehensile and the brood is mostly exposed under the tail or trunk section. The DORYRHAMPHINAE are a group of free-swimming pipefishes that have a mostly exposed brood, and a large flag-like caudal fin .

All family members are basically small and secretive fishes that feature elongated semi-flexible stiff bodies, armoured with bony plates and rings instead of scales. The gill-opening is reduced to a small round pore, the head usually has a long tubular snout with a small mouth at the tip, and the jaws lacks teeth. The length of the snout is highly variable, depending on specialisation of a particular species. Fins are all soft-rayed (lacking spinous rays), variably present or absent, and some worm-like species have none. Most species possess a single dorsal fin and pectoral fins, and in some groups there is a moderately sized caudal fin. Ventral fins or second dorsal fins are absent in all species. The anal fin is usually small or degenerated, but maybe prominent in hatchlings and planktonic stages. Fins are used for swimming and manoeu-

vring, the dorsal for going forwards or backwards and the pectorals for up and down. The fins are best developed in the free-swimming species. In the seahorse the dorsal and pectoral fins are moderately large and they propel themselves by undulating these rapidly. Whilst all species are essentially benthic, swimming behaviour varies considerably. Most species are continuously in touch with the substrate by simply laying on the bottom or clinging onto things with their tail, but one group (DORYRHAMPHINAE) prefers free-swimming close to the bottom, walls, or on ceilings of caves .

Seahorses and pipefishes range in size from tiny pygmy species, just a few centimetres long, to the largest pipefish over 65 cm in total length.

Apart from their exclusive look, these fishes share another interesting behavioural aspect: their reproductive strategy. All members of this family share a unique reproductive method in which the male gets pregnant! Instead of the male taking the sperm to the egg, for the female to incubate, the female 'smartly' deposits her eggs in a pouch (*see photograph on the left*) or special patch of skin of the male where the sperm is waiting instead. Thus, the male becomes pregnant and incubates the eggs. The incubation period is considerably long for a fish, and most species give birth to highly developed offspring. Species with prehensile tails that are continuously in physical touch with the bottom need have the best protection of their brood. This varies from a fully enclosed pouch to interlocking skin-flaps to covers the eggs completely, keeping out any crawling predators, whilst the free-swimming species offer the least protection to their brood with eggs exposed to the outside.

Mysid spp from Port Phillip Bay

All species appear to be diurnal and carnivorous, preying primarily on small crustaceans that are sucked up whole. These are located by sight and picked off the bottom or from the water column depending on species, their stage and opportunity. Species that prefer free-swimming prey, such as mysids (*see photographs on the left*) or larval fishes, lay in strategic places along the edges of reefs or seagrass beds with moderate currents where food passes close over the sand. With their perfect camouflage they wait for prey to get into striking reach. Those that hunt the

Mysids clouding the bottom under Frankston Pier, Port Phillip Bay, Victoria, Australia.

benthic prey may simply float along looking like a piece of harmless weed or slide through algae in stop-start fashion to sneak up on them. Each species has a favourite habitat that provides the optimum opportunity for food and gives them the best protection. Most are Marine and associate with particular algae or weed habitats on coastal reefs, but some have adapted to special places, live on a specific host, or have moved into freshwater systems. Apart from a few conspicuous coral reef dwellers, the species are highly camouflaged or look like part of the environment, mimicking anything from leaves, sticks to pieces of weed, depending on habitat preferences. Those living in a seagrass bed look like the seagrass, whilst those out on the open substrate may appear like pieces of loose weed or sticks. Colour can be variable within species, especially in those that live in mixed soft-bottom habitats, matching certain algae or sponges.

Despite their great camouflage and strategic moving about, the adults are preyed upon by many benthic fishes, such as flathead (Platycephalidae, see below), and various snapper groups. Once discovered by a predator, they have little chance to escape. Their body hard outer skin provides no protection to large predators who swallow them whole. Some species have pelagic young and during that period the mortality rate is very high, many falling prey to various planktivorous fishes. Sub-temperate pelagic young are often taken by seabirds such as penguins. Once adult stage is reached, the greatest danger has past and most species will probably produce numerous broods in their lifetime.

Except for a few habitat-specific species, these fishes are easily kept in captivity by the serious aquarist as long as they are provided with regular food, good water quality and appropriate surroundings. When obtaining specimens for the aquarium, it is important to know where they came from to make sure that the correct temperatures are provided and the right food is available at all times. Incorrect temperatures and poor water qualities results in loss of colour, followed by loss in condition and vulnerability to diseases that are difficult to cure. To keep these fishes successfully, it is most important to find out beforehand what not to do, as well as creating a good home. Only when all the right parameters can be met, it is justified to keep these fishes. With regards to sickness, prevention is the key to success. Poor water quality effects the sensitive brood area of the male with infections, and unhappy specimens may commonly break the surface with their snout that can attract fungal problems to the exposed areas. Curing such problems is very difficult. Badly shipped specimens usually show problems around the snout that may look bruised, and should be avoided. If such specimens are purchased, quarantine is absolutely necessary. In any case quarantine is recommended. Specimens may carry a disease to which they are immune, but may infect others that are not. This is especially important to note when specimens originate from different geographical zones .

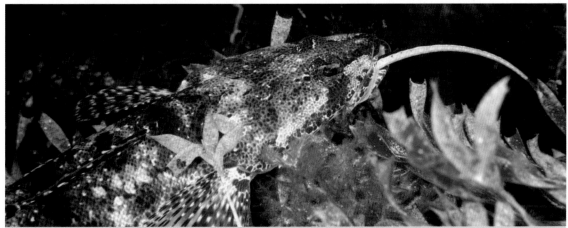

Platycephalus laevis eating a large pipefish. Flinders, Vic, Australia. Fisherman commonly find seahorses (*H. breviceps*) of various sizes in their stomach.

A universal classification of all life forms started about 200 years ago. This science is called taxonomy and the aim is to accommodate these forms in order of relationships with a common internationally language, into an organised framework. A unique life form is a species, and is referred to as a taxon. Closely related species are grouped together into a genus, a higher taxon, and similarly genera into families, etc. Those grouped together form a level, the various forms or groups are taxa.

Scientist working with systematics are called 'taxonomists'. The system is known as taxonomy. The most recently formed level is the species, a life form reproducing itself and retaining its unique identity. Closely related species that derived from the same ancestor at a point in time, are placed in a genus, similarly related genera in families, following with other levels, as going further back in time, orders, classes, phylums, and so on, that make up the kingdom.

ORDER: SYNGNATHIFORMES
SUBORDER: SYNGATHOIDEI
FAMILY: SYNGNATHIDAE **- PIPEFISHES**
SUBFAMILY: HIPPOCAMPINAE **- SEAHORSES & PYGMY PIPEHORSES**
Genus: *Amphelikturus* Parr, 1930
1. West Atlantic Pygmy-Pipehorse *Amphelikturus dendriticus* (Barbour, 1905). Bermuda.
2. East Atlantic Pygmy-Pipehorse *Amphelikturus* sp (in Dawson, 1982). East Atlantic.

Above, *Classification example as used in this book* in which the Order SYNGNATHIFORMES encompasses all the related species featured. The genus is written in *italics* and followed by the person who described this taxon and the year it was published. After the species is the name of its describer and the publishing year. These are put in brackets if the genus has been changed from its original. In this case it was described as *Siphostoma dendritica* Barbour, 1905. Since that genus is another kind of pipefish (= synonym of *Syngnathus*), this name can't be used and it was renamed by Parr as *Amphelikturus*, and because its gender is masculine the species name was altered accordingly.

When a species is not yet described it is listed as **sp**. If one refers to several, it is written as **spp**. The reference is given in brackets.

Names for the higher level has derived in principal from the first name used in the level below. E.g. the name SYNGNATHIDAE for the pipefish family derived from *Syngnathus,* the first described genus.

Species are particular life forms at the present point in time, and are the products of changing environments. Every new generation produces a range of slight variations that may offer some advantages to changing circumstances that faces populations. The best suited is the most successful that continues the line. Over time species can change differently between populations and change to species in their own right. Ancestors from a long time ago may have looked very different, and would be genetically incompatible with their present offspring, thus are now extinct. Such extinctions are the results of normal genetical drift, but not all extinctions are so kind. Extinctions can be caused by natural disasters, wiping out habitat and spelling a sudden end, or take toll on those that become too specialised and dependant on many parameters in a delicately balanced environment. Slight alterations in such sensitive habitats may inhibit reproduction, causing dying out gradually. Species may become too successful and causing others to struggle during their domination. A single coral species may cover an entire reef if not destroyed by storms on a regular basis. The human race dominates the entire planet and consumes habitats, driving countless species to extinction .

Many species have a common ancestor and looking back further, those earlier ancestors evolved from even earlier ancestors, leading eventually to a smaller and simpler pool of life. Life is like an explosion in super-slow motion, sending life forms from its origin on an evolutionary journey in all directions, in which the species are the leading bits. Early life forms were simple and few and evolution moved at a slower page compared to now, accelerating and gaining momentum. It is often compared to a tree where you have a simple trunk and branches are few and thick near the trunk, becoming more numerous, thinner and change directions more often and quicker. In a tree the direction of growth is towards the light and influenced by the surrounding branches, which is not unlike the way species are

influenced by their surroundings. An evolutionary tree would look more like a fuzzy ball, or a sphere with countless tips. The tips closest to each other, left right, above below etc are closest related. A tip on the opposite side of the sphere is a far-distant relative, but maybe equally as much evolved over time, or at the same distance from the centre of the sphere.

Evolution is a continuing process, and it is sometimes impossible to determine the exact level of a taxon. Species that are widespread may occur in semi-isolated populations, in which similar forms could be variations or species that are in the process of going separate ways. The greatest dilemma is when determining the status of species that comprise several neighbouring populations in which those next to each other are only slightly different, but the most separated ones are at different species levels. If the intermediate populations were not existing, there would be no argument, and often such species are paralleled in distribution by other sibling species from other families without the intermediates. In many families there is disagreement amongst taxonomists to where to draw the line, which in fact is not possible .

Speciation occurs when populations or faunas are divided into separate or additional geographical sections that drift apart. Faunas may move in various directions along continents, following conditions requiring the least change. Species travelling along will change only slightly from their original form, whilst those left behind adapt to the changes and become the most different species. This has resulted in species-complexes in which the most similar species are found at the extreme of their geographical range. In the Indo-West Pacific there are three near identical faunas identified between eastern Australia, southern Japan, and eastern Africa. Isolated populations undergo evolutionary changes that differ from each other, depending on time versus environmental pressures. The most recently isolated populations may look slightly different, but may interbreed when drifted back together, and such forms are usually referred to as subspecies or geographical variations. Once they can not interbreed, even if hybrids are produced (normally these are infertile), they are regarded as separate or valid species. In general, similar species that are geographically close together are more recent than those further apart.

The rate of change in a species is determined by a combination of many factors. There is a natural genetic drift that is relatively slow, but habitat alterations demand adaptations accordingly. Competition with similar species or preying pressures may speed up change in a species at much faster rates compared to the same species elsewhere, without the additional pressures. Over time, isolated populations become true species.

In highly diverse groups such as the pipefishes and seahorses, the family is divided into several subfamilies, genera into subgenera, and species are often grouped as tribes within the subgenera. Few people have worked on these particular fishes and it is clear that the taxonomy based only on morphology failed to recognise many species.

As most seahorses and their relatives have highly developed offspring, their distribution is generally restricted, but some tropical species are drifters in loose weed rafts, including as adult, and these are widespread. Presently all seahorses are placed in one genus *Hippocampus* but there are some differences between tropical and subtemperate species and further studies may warrant subgenera. The pipefishes show much more diversity and consist of many genera.

Fossilised ancestors of pipefishes and their relatives, from around 50 million years old, are known from Europe, North America, Africa and west Asian continents. They represent forms that are very similar to the more elaborate family members today. Some of these, such as the seahorse and some pipefish, are thought to represent the same genera.

The presently suggested relationships with the other families may seem questionable when only looking at recent species. Such relationships become clearer when fossils are included in the studies of the families, as these are more primitive and less modified than recent species. In fossils the various features that were common between different groups, may have degenerated in modern species such as certain fins. Some extinct forms represented links between different living groups.

NAMES & SYNONYMS

Fishes are given scientific names when described by Ichthyologists: researchers that usually specialise in particular kinds. Names are expressed in Latin or it originates from classic Greek and usually refers to a particular feature that is of significance to the species. It may also reflect a place of capture, a land mark or is named after a person. Species names are binominal, = in two parts, first is a genus and second the species, and are also of a gender: masculine, feminine or neutral. The species is of the same gender of the genus and the difference is expressed in the ending of the name. Many masculine species names end in '*us*' of which the feminine equivalent is '*a*'. Scientific names are normally written in Italics, the genus starts with a capital letter and the species always in lower case. Eg: The Estuary Seahorse *Hippocampus kuda*, in which *kuda* is the species and *Hippocampus* the genus. When names are repeated in text, the genus is abbreviated and species written as *H. kuda*. Other seahorses that belong in the same genus share the genus name, such as the Thorny Seahorse *Hippocampus histrix*, or *H. histrix*. It's like people having a Christian and family name, but family name comes first.

Descriptions are based on one (holotype) or more specimens (types) and published in journals of institutes involved in fish taxonomy, whilst the type material is housed in one or more institutions. The taxonomy of fishes is relatively young, and there are many similar species that were only recently discovered. The nomenclature includes numerous synonyms and wrong names are often applied, even to the most common species, and only when a particular group is studied in detail, this becomes apparent. Many problems came about when similar species were treated under the same name and later were recognised as being different, but then applying the name to the wrong one. A popular way of determining a species in similar groups is providing a key that is based on the characteristics that separates one species from the other. However, this can also be a trap as such a key may not cover all the species and an unknown or new species may fit the key. Wrong names often result from the use of keys to determine a species when used in different geographical zones that often share closely related species. When an unknown species fits the key it receives the wrong name, and this is a common problem in general surveys. EG the seahorse *H. whitei* that is only found on the eastern coast of Australia was used in the South African literature because the similar *H. borboriensis* fits a key. Keys are very useful to determine genera and in areas where diversity at species-level is not great, such as in temperate zones or the

Atlantic. Unless already an expert, keys are of little or no use to work to species level in the tropical Indo-West Pacific.

The discovery of new species can also lead to questioning the name used for another species as it maybe a better match to the original description. In many cases confusion remains about the identity of species that were named a long time ago. Descriptions and illustrations were excellent by many authors and with diagnostic features presented leave no doubt to their identity, but others were basic and poor, and could apply to several species. Type-materials that represents particular species may be lost, was destroyed in wars or natural disasters, or in some cases specimens were substituted with others that may not represent the true species. When there is a problem with the identity of a species, all these factors need to be considered. As many species have a restricted distribution, the type locality can play an important role to determine the true identity of a species.

Confusion is caused when the same species is described several times by different people, different forms of the same species received separate names, or the same name is used for different species. Many species were described over 100 years ago and at the time it was difficult to know if somebody else already named the same species. This often resulted in common widespread species having multiple names. The first name and description has priority and accepted as the correct scientific name, and is the senior synonym. All others that came later are called junior synonyms. A name in use may become invalid and replaced when an older name is found. Similarly generic changes can occur, or species are assigned to other genera as relationships are clarified. When a species is changed to another genus that is of a different gender, the species-name also changes gender. Rules are determined by an International Commission but, like evolution, change over time, basically to make things run smoother in the system and iron out conflicts. For instance, some of the original methods of writing names was changed, such as species that were named after persons started with a capital letter, or letters that were not international have been changed. Thus, the original species name of *Güntheri* is now *guentheri*. Bleeker, a Dutch Ichthyologist, described numerous species of the Indo-Pacific using the Dutch "ij" (single letter, not i+j, and joined when hand-written, but typed with two letters, simply because it is absent on the machine) and is equivalent of "y" in other languages, which has led to some confusion. Eg for *polijtaenia* the correct spelling is *polytaenia*.

GENERAL

The purpose of this book is to serve as a pictorial guide with a comprehensive coverage of the family SYNGNATHIDAE with as many species as possible, including the popular seahorses, seadragons, pipehorses and pipefishes. The closely related families, SOLENOSTOMIDAE and superficially similar PEGASIDAE, are also included, and families that show affinities with them are represented with some examples. The seahorses are probably the most advanced group in the family and are treated first, followed by the pipehorses, and ending with the simpler pipefishes. This family is followed by the related groups, which are similarly arranged. Information covers all known aspects of the included species, such as ecological & geographical distribution, behaviour, biology, aquarium suitability or care. It also serves as an identification guide for fish watchers and aquarists, showing the species with numerous colour photographs, variations, different sexes, as well as many stages of development. The species are grouped together in their genus and are arranged in order of similarity. This may differ from scientific order that usually starts with the most primitive species and ends with the most advanced.

Each species is given a common and scientific name. The common name can be in various languages, according to the country or locality, whilst the scientific name is international. Established common names are used as much as possible and where several are known for a single taxon, the oldest or most appropriate is chosen. As the snout length varies greatly in the pipefish family, names such as short- to long-snout pipefish is applied in many countries for different species. In such cases I've tried using geographical attributes such as Atlantic or Pacific when possible. This is only for the global perspective and doesn't alter the locally used name. Usually the name reflects a unique feature of the species or relates to habitat, locality, or is derived from the scientific name. Species that were lacking names and are published here for the first time are given new common names in the same way.

Notes on biology, ecology and any interesting aspect of a genus or species are including in the text of the appropriate taxa. Notes on aquarium associated subjects, including requirements, diseases and prevention, are either in the genera or species text, depending on the extend to what is known.

TERMINOLOGY

To avoid any confusion, many terms used to describe the seas or areas in this book need to be explained. They can mean different things to different people or in different countries, and often are seen from a certain perspective. Eg: For someone living in Greenland, most of the areas included in this book seem tropical, whilst on the other hand an Indonesian diver would freeze in Sydney's or Tokyo's waters.

The Seas.

The upper layers of the oceans around the world are warmed by the atmosphere and have a broad range of temperatures that are in relation to its geography and currents. Gravity forces and the rotation of the planet pushes the water masses between continents in a clockwise and anti-clockwise direction, above and below the equator respectively, and may bring cold up-welling to certain areas. The warmest waters are near the equator and coldest near the poles. Currents even out some differences, and there are many principal ocean currents running in set patterns, that have been identified and named. However, such currents can be highly complex and consist of eddies that causes local variations with an influx of tides. There is a some degree of instability and

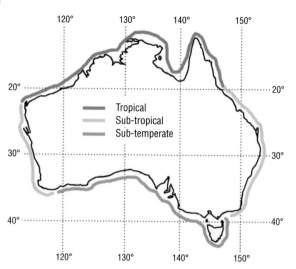

MAP 1. Bioregions in Australia, based on distribution of pelagic fishes. This reflects the broad distribution of the different faunas. Demersal faunas are much more limited and can be restricted to small areas, such as the gulfs in South Australia, or an area such as Bass Strait. Most species in SYNGNATHIDAE do not range over any of the entire regions, some are close in their range to one of the regions, and few range into other neighbouring ones. The habitat-specific species may have a sporadic distribution and usually have isolated populations. Some species are only known from small estuaries and require monitoring.

seasonal changes in the overall patterns, which are often cycling on different time scales and can range from months to many years. Light and temperatures are the foundations of faunas and tropical reefs look very different from those in cold water. There are some places where one can see changes from one kind of fauna to another and the zones that support particular faunas are termed biogeographical zones. The most different biogeographical zones are between the seas with various temperatures, ranging from the cold-temperate to tropical, from the poles to the equator respectively. Overlapping areas are less defined and may fluctuate greatly with seasons and currents.

Tropical

Most people can relate to *tropical*, the comfortable warm belt around the earth in vicinity of the equator. The warmest waters, and to be more specific, tropical waters refer to seas that have annual temperatures of 20° Celsius or above. On maps with sea-temperatures this is normally the red zone.

Subtropical

The zone that borders the tropical seas towards temperate zones, between tropical and subtemperate. Annual temperatures range from about 20° down to 12° Celsius. Maps usually show these zones in a range from orange to yellow, warm to cool respectively.

Temperate

The coldest seas, annual temperatures usually less then 5.5° Celsius. Maps usually show a dark-blue for this zone.

Subtemperate

The zone that borders temperate seas towards tropical zones, between temperate and subtropical. Annual temperatures range between about 5.5° and 12° Celsius. Maps usually show these zones in a range from blue to pale blue or yellow, cold to warm respectively.

Fairy Penguin *Eudyptula minor* takes lots of small pelagic seahorses in the Bass Strait region of southern Australia.

as some seahorses love man-made structures, such as jetties, that naturally are built in suitable habitats and object that include ropes and shopping trolleys. Jetties usually reach where the bottom slopes to deeper water, often where tidal currents run that carry planktonic food. Mussel farms in open estuaries are perfect for seahorses, as these are usually in areas with plenty of food but otherwise provide little shelter. The ropes used are ideal to wrap their tail around, and this has led to a dramatic increase of seahorse populations there. In most places, seahorses are too costly to collect from the wild for dried material, especially by diving with a breathing apparatus. The largest species are in subtemperate waters and these disperse over large areas in moderate depths. In addition the habitats are difficult to reach because of distance from shore without boat, as well as land travel to get to sites. In developed countries the cost of transport and diving would far outweigh the return for a catch. The good news is that the environment is becoming much more of an issue and part of the education in schools. Replanting and fixing waterways is much more realistic to protect the seahorse than making it illegal to collect one without a special license, which is now the case in Australia.

Seahorses are most vulnerable in their early stages, especially those that have pelagic stages. A study on Fairy Penguins (see photograph above) in Victoria, found many specimens of juvenile *Hippocampus breviceps* in their stomach (they were identified as *H. whitei*). In some areas, fishermen find adult seahorses often in the stomachs of fishes they caught. In Western Port, Victoria, flathead (Platycephalidae) regularly have *H. breviceps* in their stomachs. In Sydney, NSW, the Striped Anglerfish *Antennarius striatus* feeds on *H.* cf *abdominalis* and *H. whitei* and several divers have observed them stalking their prey and swallowing them (see photograph on right).

All seahorses are presently placed in the single genus *Hippocampus*. Whilst the various species can be placed in different groups, that share certain behaviour or morphological variations, no attempt has been made here to reclassify them. The differences appear to be well short

Anglerfish *Antennarius striatus* eating *Hippocampus abdominalis*.

of creating additional genera, but future studies may show a case for dividing the genus into a number of subgenera. A fossilised seahorse, about 7 million years old from Italy, shows how little change, if any, has occurred over such a long time. There are about 110 seahorses-descriptions, but often the same species was named by different workers. About 70 are now recognised as valid species, plus there are a small number known that are not yet named, and more can be expected. For many species the identity and synonyms are clear, but for some others it is almost impossible to be sure. Some early descriptions were vague, the type material missing or badly damaged. If in such cases several nominal species can be found in a type locality, the applied name may remain questionable. Established early names, such as *H. kuda,* are widely applied to various species.

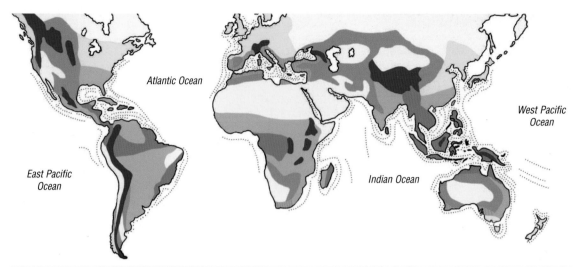

MAP 3. Distribution of seahorses around the world. Approximate distribution represented by dotted lines. The majority of the species in the family SYNGNATHIDAE fall within the same range, but are marginally more broadly distributed into sub-temperate or oceanic regions.

DISTRIBUTION

Seahorses are found in all but the cold temperate seas around the world. Most species live in tropical and subtropical zones, but few can be found the subtemperate zones of Australia, New Zealand, Japan, South Africa and Europe. None of the species have a broad geographical range. Literature records suggest that some species have a broad distribution, but this is now clearly wrong. All species are restricted in their distribution to a specific biological region, even in the tropical zones such as south east Asia, which comprises a number of different faunas that have evolved independently. In comparable faunas between different regions of the world, the species may be so similar in morphology that they can be difficult to distinguish when preserved. However, in life colour can be very different and their habitat requirements are rarely the same. Usually species change very slowly in morphology over time and some of the present species are near identical to fossils. Changes are apparent in arrangement of spines, proportional changes of shape, maximum size and colour. In regions where seahorses occur there are usually at least two species, one that is a shallow sub-littoral dweller, and one that has adapted to soft-bottom habitat, a slightly deeper zone. Faunas that have separated through continental drift share many similar species that are siblings. Such regions are identified in the Indo-Pacific between Japan, Australia and South Africa. Similar seahorses occur in these regions that have a very similar morphology and consequently names have been exchanged between those regions in error.

Seahorses are well developed when born and many species have no pelagic stage. Few can be found drifting on the surface in loose weed, including when adult, but it seems that their movement is localised to within a certain region by tides. In tropical seas, some species are washed out of estuaries with loose weed during the wet season that primarily serves to escape the influx of freshwater that they can't tolerate, and sometimes are found far offshore. Migration is known for a number of species that occurs in estuaries and they build up in numbers to breed after the wet season. Food becomes abundant, as many crustaceans use estuaries to breed and their larval stages or young are the main food source for many juvenile fishes, including seahorses. Some species move to other places after breeding because of food becoming scarce. Males may prefer different habitat and different food from females, and they are occasionally found in small groups of the same sex.

Generally, most seahorses are reef-fishes that are habitat-specific, preferring algal or sponge reefs in 5 to 30 m depth, depending on the species or area. Sometimes species are found around the edges of shallow seagrass beds and others may live deeper than 30 m on soft-bottom substrates. Some subtemperate species may breed in the shallows and found deep during other times. There is a misconception that all seahorses associate strongly with seagrasses, but this is a reflection of divers having easy access to them in shallow depth, or from collecting or trawling. In addition, seagrass beds are often used for research projects. Seahorses are more difficult to find in deeper water and trawlers avoid reef habitats where most are found. Few species have associations with soft corals to which they can adapt in colour to perfection by supporting some of the living coral tissue in their skin.

The 60 or so different seahorses are variously distributed in the tropical to sub-tropical or sub-temperate seas around the world, and new species are still being discovered. Some are only known from a few specimens that were trawled in relatively deep water. Seahorses are difficult to find, including the most common ones, due to their perfect camouflage and discovery is often accidental. Although large numbers of seahorses are collected in south east Asia for medicinal purposes, relatively few species are exploited. These live in a variety of habitats with a moderately large geographical range. Collecting is usually from shallow lagoons near villages and the species remain numerous in many other areas that are not fished, or live in slightly deeper water as well. Some of the sub-tropical or warm-temperate ones are endemic to small geographical regions and usually of no interest to collectors. Trade in dried seahorses, other than medicinal purposes, should not be encouraged by buying them as souvenirs. Calls for total protection, especially collecting from the wild, is unwarranted and counter productive. People who want to learn more about them are best served by keeping them as pets. Habitat alterations are the real threat for seahorses at present. Species which have a restricted distribution, and only live in shallow coastal regions, could become vulnerable if habitats are altered. Coastal development near estuary channels, seagrass beds or mangroves has an immediate effect, causing turbulence and loss of vegetation. Habitat for many small species, and food-source for many juvenile fishes is lost. It is unlikely that seahorses will become extinct from collecting them. Loss of habitat and pollution is the greatest worry to the possible decline in some populations.

Diver observing courting behaviour of *Hippocampus bleekeri*, Rye Pier, Port Phillip Bay, Victoria, Australia.

SEAHORSES AND CAPTIVITY

The picture on the left shows a diver taking an interest in seahorses. Seeing any creature in its natural environment is the best way, but unfortunately few people get the opportunity to do so in the ocean, and the aquarium is the next best thing, either the public aquarium or at home. There is still much to learn about seahorses. A divers time is limited and the general behaviour of seahorses, besides feeding, is usually on sunrise. Many species occur abundantly in the wild and are easily kept, and people should be encouraged to take an interest in these special fishes. Each species needs to be studied in its own right and until now only few species have been studied in detail. Most studies have been selective and involved few species. Findings are often generalised and may wrongly suggest to apply to other species. Claims of *monogamy* in seahorses either misinterpreted or overstated and I have avoid using this term, as these fishes simply pair-up like many other fishes do, and the level of devotion varies from strong-bonded to casual, depending on the species, and partner swapping occurs in all at some stage. Information on the species, based on my own experience, are given in the species account and are most detailed on Australian species. Little information has been published on species elsewhere, and most are done by hobbyists. I would like to hear from anybody who likes to contribute to the next edition on the seahorses.

Sincerely,

Rudie H. Kuiter

HUSBANDRY

GENERAL

Mostly the common and larger species have been kept in captivity, and all have proven to be easy and readily breed in a healthy set-up. Success is a combination of devotion to the creatures, to give them plenty of space, and a surrounding that resembles their natural environment. The availability of correct food at all times is essential to maintain the good health of the specimens. Poor diet results in colour loss, and worse, infections that usually develop on the more sensitive areas such as the snout, and the pouch of the males.

OBTAINING SPECIMENS

When purchasing a seahorse it should show no damage on the snout or have buoyancy problems. Identify the species. Temperatures should closely match the sea temperature where the specimen originated from. Subtropical species may require cooling systems and should not be tried in tropical aquariums. Tropical species can handle lower temperatures better than cooler water species can handle warmer or higher temperatures. If in doubt, don't try. It is important to quarantine new arrivals before introducing it with other seahorses or pipefishes. Keep it (or them) separate for several weeks and make sure feeding is normal as well as checking for signs of trouble such as fungal development around the snout or skin. Treatment of diseases can be very difficult and is often fatal. Their special gill structure makes them sensitive to chemicals and they can only handle mild doses. Dropping salinity is often the first and best option. Catching your own specimens is ideal as one can look for paired specimens. It is important not to introduce these into an established aquarium with other seahorses or relatives that have been kept for a while, as these may have become sensitive for wild parasites or diseases. Always quarantine them first.

FOOD

They are diurnal feeders and spend all day in search of small crustacean type foods that can be sucked up from rocks or weeds, as well as planktonic creatures that float passed into reach. Ideally living mysids are supplied that can live in the system, but any small shrimp or prawn can be fed. Some species like to attack shrimps too large to be swallowed whole and manage to suck the abdomen off, leaving the rest, or can be trained to take baby fish that can be bred, such as *Tilapia*. A variety of small crustaceans and availability for most of the day is essential to breed species over generations and maintaining a strong line. A commonly used food is brineshrimp, especially when raising young, but is a poor substitute and needs to be enriched, but the success rate of raising young is generally inferior, compared to those raised on mysids or shrimps. In addition the descending generations become less productive and produce smaller broods of weak young. However, brineshrimp are useful as a substitute when there is a shortage of better food. Seahorses should be kept only with other small fishes that are not too competitive for food. They are ideal for invertebrate aquariums but anemones, stinging corals, hydroids, or some crustaceans with large nippers can be a threat. Hatchlings are best put in a separate aquarium that has no invertebrates at all, except of course microscopic ones for food.

AQUARIUM

The aquarium should be as large as possible, the larger the species, the more space is needed. Height is important for breeding. Whilst small species are quite happy in a small aquarium, it limits the number that can be kept and most do better when several pairs can be kept, which is much more interesting. Estuary species are tolerant to temperature and salinity changes, but a stable environment is essential for breeding them and avoiding problems with disease. A large volume of water plus a reasonable stable ambient temperature are important factors. Overheating causes the biggest problems with the subtropical species that usually do best in temperatures below 20°C.

LIGHT

Light can play an important roll in their well being, especially when breeding seahorses. Ideally the light is the natural ambient light that starts and end the day. Artificially lighting should not start or end the day, unless it is done with a slow rise and fall dimmer control. Most species go through certain behaviour at that time, including simple greeting, courting and mating. Some species are more tolerant than others and those that naturally live deep are most sensitive. Certain species prefer a low light level at the time.

Comprises 1 genus of seahorses and 3 genera of pygmy pipehorses.

GENUS *HIPPOCAMPUS* Rafinesque, 1810

Masculine. Type species: *Syngnathus hippocampus* Linnaeus, 1758. The seahorses, about 50 species globally distributed in tropical and sub-temperate seas. See page 10 for details.

White's Seahorse *Hippocampus whitei*

Hippocampus whitei Bleeker, 1855. Sydney. Australia.

A common New South Wales species in the large estuaries of the Sydney region, ranging north to Forster. Reports of *H. whitei* from southern states need to be verified as some were based on young *H. breviceps*. *H. whitei* occurs in a variety of sponge-reef habitats to moderate depths, and mixed reef and seagrasses in sheltered coves. They may breed throughout the year but are most active in spring and early summer, pairing in selected areas. Most pairs stay together and may have several broods per season. Mating and birth activities peak on a monthly cycle when the largest tides eventuate near full moon. Several pairs in captivity kept in phase, producing about 60–80 young per brood per month. Young don't have a pelagic stage. All but few were released and some grown to adults. After about 3 months sex could be determined as a pouch showed clearly on the male, and within 6 months they are ready to breed. In the wild the winter period may have deterred them until spring and by then would have attained full adult size. Single adults are commonly found in tidal channels of large estuaries with solitary, stringy sponges during winter months, usually in depths less than 20 m. Height to about 10 cm.

A　*H. whitei*. Sydney Harbour, Australia. Height 10 cm, giving birth.

B　*H. whitei*. Sydney Harbour, Australia. Height 9 cm.

C　*H. whitei*. Sydney Harbour, Australia. Height 9 cm.

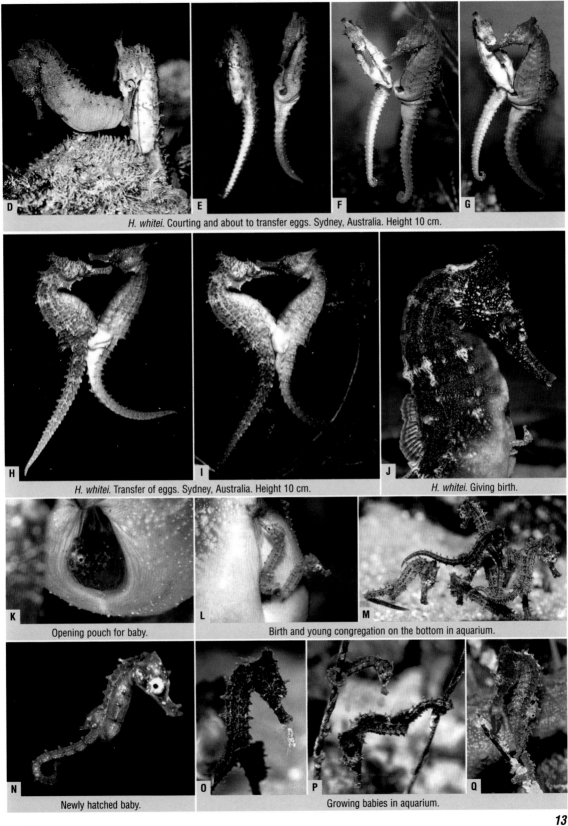

H. whitei. Courting and about to transfer eggs. Sydney, Australia. Height 10 cm.

H. whitei. Transfer of eggs. Sydney, Australia. Height 10 cm.

H. whitei. Giving birth.

Opening pouch for baby.

Birth and young congregation on the bottom in aquarium.

Newly hatched baby.

Growing babies in aquarium.

High-crown Seahorse *Hippocampus procerus*

Hippocampus procerus Kuiter, 2001. Fraser Island, Qld.

A recently described species that occurs in coastal Queensland from Moreton Bay to Gulf of Carpentaria. Was confused with *Hippocampus whitei* and reports of this species from PNG and the Solomon Islands are probably based on *H. procerus*. It differs in various ways from *H. whitei*, most obvious is the taller and spinier crown, but also has more fin rays and sharp spines on the ridges. It occurs on shallow algae reef and rubble substrates to about 20 m depth. Height to about 12 cm.

H. procerus. **A** Male, height 13 cm. **B** Female, height 12 cm. Moreton Bay, Qld.

H. procerus. Small male, height 9 cm, courting female, height 12 cm. Moreton Bay, Qld.

H. procerus. **D** male. Height 13 cm. **E** female. Height 12 cm. Moreton Bay, Qld.

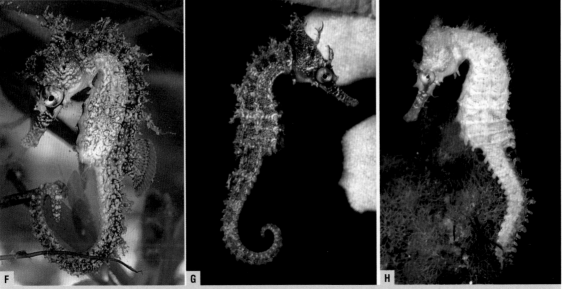

H. procerus, juveniles. **F** height 8 cm. Bruce COWELL. **G** & **H** heights 9 & 10 cm Moreton Bay, Qld.

Hippocampus breviceps Peters, 1870. Adelaide, S.A.

Australia's south coast, St Vincent Gulf to Bass Strait region. Various weedy habitats, often in sargassum along edges of seagrasses. Sometimes on sponge reef in deeper water. Usually with long appendages on the head and back when in sargassum weeds. Juveniles with long snout, 1/2 head length, shortening in adults. Specimens settling on the substrate at later age may develop longer snout compared to those settling earlier. Occurs in small groups that can be found congregating in safe places at night, usually high in the weeds to keep away from the crabs on the substrate (**A & I**). Feeds close to the sand or rubble during the day, usually targeting mysids. Found from sub-tidal to about 15 m depth, but may venture deeper. Breeds in summer and has about 50 to 100 eggs in a brood. Height to 10 cm.

H. breviceps. Portsea, Vic. Grouping together at night, only 2nd from left is female.

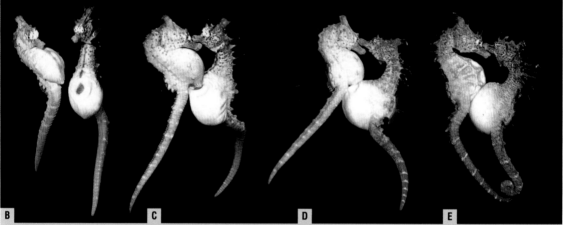

H. breviceps. Female, in each picture on the left, **B** ready to deposit eggs in the males open pouch, showing ovipositor. **C** Placing ovipositor over the pouch opening. **D** Start of egg flow. **E** Egg flow completed, female shows concave trunk. **B-D** are in sequence, **E** is a different pair.

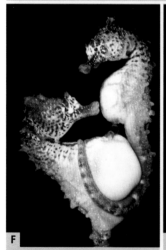

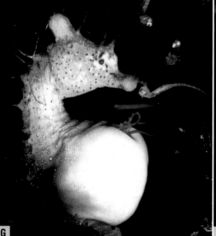

Female anxiously departs, exposing the last two eggs, the male managed to secure them afterwards.

H. breviceps. Male giving birth. Young are born after about a 25 day incubation time. Young appear one at the time or in batches, sometimes few, and other times lots, probably depending on the young working their way out of the egg. Birth may last several hours but usually is much shorter. Young swim to the surface.

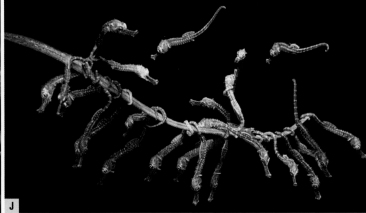

J

H. breviceps. Hatchlings cling to each other until something better comes along. During the summer months, large numbers of young cling to bits of weed in surface waters, and are often seen from boats on the large outgoing tides near full moon. Many will fall prey to other fishes, but also are part of the diet of the Fairy Penguin *Eudyptula minor* (see P. 9).

I

H. breviceps. Moonta, S.A. Night, males.

K

H. breviceps. Off Rye, Vic. Pelagic young about 25 mm height, near settling size.

Knobby Seahorse
Hippocampus tuberculatus

Hippocampus tuberculatus Castelnau, 1875. Swan River, Perth, W.A.

Western Australia, Perth region to Onslow. Brownish black to orange or reddish in general colour. Juveniles to sub-adults with long white area between eyes and from below coronet to tip of snout. Adults often colourful with red tubercles, depending on habitat. Juveniles and subadults, in which males have fully developed pouches, often float off-shore in sargassum weeds, but older ones are usually on sponge reefs in depths of about 20 m. It was previously confused with *H. breviceps*, that occurs in cooler temperate zones, but it has different characteristics and is much smaller when fully grown. Height to 55 mm.

A

H. tuberculatus. Female, height 50 mm. Shark Bay, WA. From 20 m depth. G.R. ALLEN.

B

H. tuberculatus. Off Perth, WA. Sub-adults in floating Sargassum weed. Clay BRYCE.

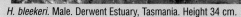

Australian Pot-belly Seahorse *Hippocampus bleekeri*

Hippocampus (Macleayina) bleekeri Fowler, 1908. Victoria, Australia.

Southern Australia from Lakes Entrance to the Great Australian Bight and Tasmania. Colours range from drab grey and brown to bright yellow, and orange depending on habitat, depth and location. Their habitats range from shallow *Ecklonia* reefs to very deep sponge, soft-bottom substrates. Males and females often congregate separately, but sometimes found in pairs during summer months. A gravid female may attract a number of males which compete by inflating their pouch to entice the female to deposit her eggs. A productive species that produces over 400 eggs per brood when fully grown and has been kept in captivity for over 7 years and probably has a natural life time of about 10 years. When conditions are good they have population 'explosions' and become abundant at times in places such as Port Phillip Bay and Western Port. A large species that can reach a height of 35 cm in the colder southern part of its geographical range.

Very easy to maintain and breed in captivity. This species is exported to numerous public aquariums around the world, preferring it over their local species because of its large size as well as being ease to reproduce. It needs lots of space and to breed it successfully the tank-height preferably more that 1 m.

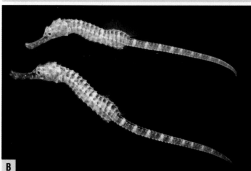

A *H. bleekeri*. Male. Derwent Estuary, Tasmania. Height 34 cm.

H. bleekeri. Port Phillip Bay, Vic, Australia. **B** Pelagic young ~55 mm long, off Rye. **C** male, height 20 cm. **D** female 'hiding', height 22 cm.

H. bleekeri. Port Phillip Bay, Vic, Australia. Male, height 25 cm.

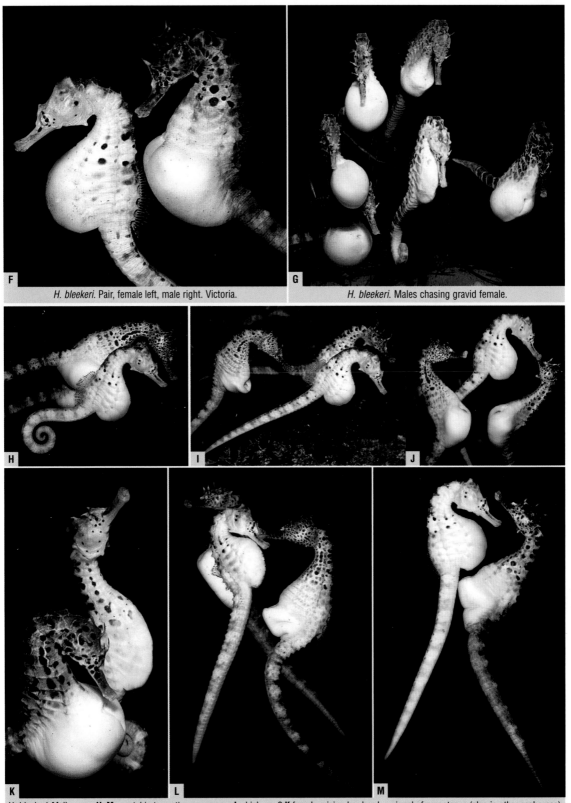

H. bleekeri. Pair, female left, male right. Victoria.

H. bleekeri. Males chasing gravid female.

H. bleekeri. Melbourne. **H–M** courtship to mating sequence. **J** which one? **K** female raising her head, a signal of acceptance (also in other seahorses).

Undetermined species.

Eastern Australia from Newcastle to Eden, NSW and probably ranges into Victoria. Presently classified as *Hippocampus abdominalis* described from New Zealand. The east-coast form lacks a nape spine and has much lower ridges on the body and tail rings compared to the New Zealand one. Occurs in large estuaries on shallow low reef with *Ecklonia*-kelp to sponge habitats where usually in depths greater than 20 m. Colour varies from dark brown to yellow in the shallows but is more variable on sponges in deep water. Some pupil-sized spots are often present and few filaments are often present on the head of young but rarely persists in adults. Requires good conditions and regular feeding in captivity or it quickly looses colour and male develop pouch problems. Not a species for the beginner. Height to 18 cm.

H. cf *abdominalis*. Bermagui, NSW, Australia. Female, height 14 cm.

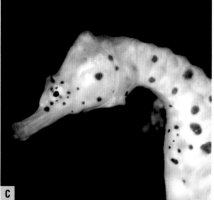

H. cf *abdominalis*. **B** Sydney, NSW, Australia. Female, height 16 cm. **C** aquarium, southern NSW. Female, showing smooth nape.

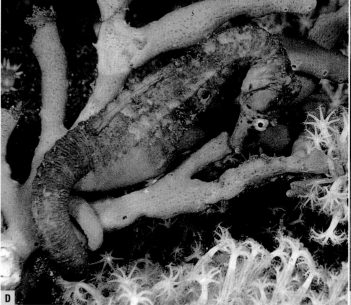

H. cf *abdominalis*. **D** Sydney, NSW, Australia. Male, height 18 cm. **E** aquarium, southern NSW. Female, showing loss of colour.

New Zealand Pot-belly Seahorse *Hippocampus abdominalis*

Hippocampus abdominalis Lesson, 1827. New Zealand.

A probable endemic species to New Zealand where widespread, but there are some morphologic and colour variations between some populations. Perhaps more than one species is present. Females with deep trunk when adult, young and males much more slender. Snout short, usually about equal to length of operculum but snout more produced in some large individuals or different populations. Colour varies from near black to yellow and orange, depending on depth and habitat. The body and tail is generally plain, often with broad banding on the latter, and the dorsal fin with numerous small spots. Habitat varies from sheltered harbours in silty open substrates to algae reef, to deep off shore sponge beds, at least to 50 m depth. Height to 30 cm.

A
H. abdominalis? Stewart I, south New Zealand. Malcolm FRANCIS.

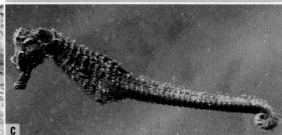

B C
H. abdominalis. Young male. Scorching Bay, Wellington, New Zealand. Malcolm FRANCIS.

D
H. abdominalis. Female. Scorching Bay, Whangateau, northern New Zealand. Malcolm FRANCIS.

E
H. abdominalis. Female. Whangateau Harbour, northern New Zealand. Malcolm FRANCIS.

Hairy Seahorse *Hippocampus guttulatus*

Hippocampus guttulatus Cuvier, 1829. French Mediterranean.

Mediterranean seas, Black Sea, and adjacent eastern Atlantic, but some variations. Found primarily shallow in algae habitats inshore, and probably drifts with floating weeds at times that contributes to their relatively broad geographical distribution. Colour variable to the environment, black to brown or yellow. Often with numerous white spots, many of which form lines across the trunk. Sometimes with numerous filaments. Breeding is typically in spring and summer months but ranges from April to October. Incubation period is about 4 weeks. Males compete to mate with females and may group around gravid ones that are ready to spawn (**A**). Height to 14 cm.

H. guttulatus. Golfe du Lion, France. **A** pregnant male. **B** hatchling. Patrick LOUISY.

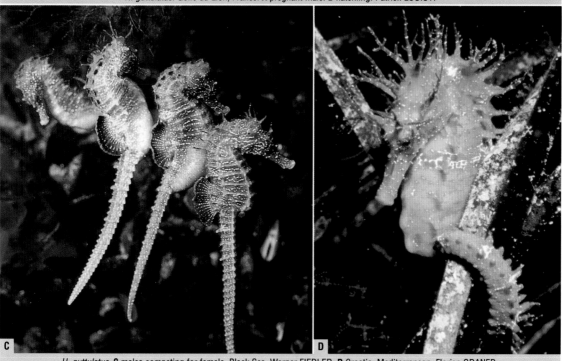

H. guttulatus. **C** males competing for female. Black Sea. Werner FIEDLER. **D** Croatia, Mediterranean. Florian GRANER.

H. guttulatus. **E** male. Italy. Thomas PAULUS. **F & G** Females. Black Sea. Werner FIEDLER. **H** female Spain. Helmut DEBELIUS

H. guttulatus. **I** gravid female pursued by keen males, one along each of her sides. **J** female in front with a male showing interest. Males usually inflate their pouch to impress the female. **K** after female chooses mate, pair rises from the bottom to get in the position to mate. **L** male opens the pouch and positions it below the ovipositor. **M** eggs are quickly transferred to the pouch that as it swells, the female's trunk shrinks. **N** males that missed out, desperately trying each other for eggs! Bulgarian coast, Black Sea. Depth 3–4 m. Werner FIEDLER.

Mediterranean Seahorse
Hippocampus hippocampus

Hippocampus hippocampus Linnaeus, 1758. Greece.

Primarily in the Mediterranean Sea and Black sea, but ranges into the adjacent waters of the Atlantic and North Sea. A moderate spine above the eye. Sometimes with sparse, but large filaments on the head and back. Dorsal fin with 16–18 rays and snout rather short. Trunk deep and ventrally rounded in females. Very similar to *Hippocampus guttulatus*, but latter has more dorsal fin rays, (19 or more), longer snout and females with less trunk depth. *H. hippocampus* can be found at various depths on algae reefs to at least 30 m, although most are observed when they congregate in shallow protected habitats during the reproductive season. Variable in colour with habitat and depth. Grows to a moderate size, height to about 12 cm.

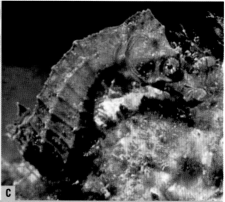

H. hippocampus. **A & B** males. Golfe du Lion, Spain. Patrick LOUISY. **C** Oosterschelde, Holland. René LIPMAN.

H. hippocampus. **D** female. Spain, Mediterranean. Werner FIEDLER. **E** male. Azores, Sta. Maria. Phil HEEMSTRA.

Europe's Seahorse *Hippocampus ramulosus*

Hippocampus ramulosus Leach, 1814. 'Europe'.

Known from the British Islands and probably in adjacent waters of Europe if suitable habitat available. Lives in shallow protected waters in sea weed or seagrass habitats. This species is distinguished from the sympatric *Hippocampus hippocampus* in having a spiny coronet and also in being more spiny on the trunk and tail. The spines often have fleshy tentacles. Colour various from dark greenish to brown, usually with with spots or blotches. Height to about 14 cm.

Remarks: it appears that this species is rare. Trying to match the type with a more common Mediterranean *Hippocampus guttulatus* led to believe that *H. ramulosus* was not a European species because of the spiny characters on the head.

A　　**B**

H. ramulosus. Cornwall, South East England. Depth 6 m. Height about 12 cm. Lee BRYANT (c/o Will & Demelza Postlethwaite).

Patagonian Seahorse *Hippocampus* cf *hippocampus*

Undetermined species, appears to be undescribed.

Mainly known from San Antonio Bay, Patagonia, Argentina. Inhabits shallow seagrass habitats influenced by tidal currents. Most common in channels during summer. This species is very similar to the European *Hippocampus hippocampus*, sharing head shape and features and dorsal fin colour pattern, and is probably its closest relative. It grows to a similar size, height about 12 cm.

A　　**B**

H. cf *hippocampus*. San Antonio Bay, Argentina. **A** female. Martin BRUNELLA. **B** male. Raúl A. GONZÁLEZ.

Lined Seahorse *Hippocampus erectus*

Hippocampus erectus Perry, 1810. American Seas.

West Atlantic, reported from the whole east coast of the United States of America, Gulf of Mexico to Florida, but few records from West Indies. The description of Perry, giving a large size, and the deep body illustrated can only apply to this species. Reported from shallow seagrass to deep water, soft-bottom habitats to over 70 m depth, and sometimes in drifting sargassum weeds. Brood size to about 400 eggs in large adults. Few seahorses are found in such a great variety of habitats and geographical range and this needs further investigation. Height to 15 cm.

A — *H. erectus*. Male. Aquarium, Scott MICHAEL.

B — *H. erectus*. Juvenile. Aquarium, Scott MICHAEL.

C — *H. erectus*. Male. Aquarium, Scott MICHAEL.

D — *H. erectus*. Female, Key West, Florida. Paul HUMANN.

Brazilian Seahorse *Hippocampus villosus*

Hippocampus villosus Günther, 1880.
Off Bahia, Brazil.

Uncertain distribution, probably restricted to eastern Brazil. Has moderately large coronet and is more spiny compared to *Hippocampus* cf *reidi*, the other species in the area. Little is known about South American Seahorses and additional photographs are required, and any information would be appreciated. Height to about 14 cm.

Remarks: previously included with *H. erectus*.

H. villosus Female. Salvador, Brazil. Depth 6 m. Bertran M. FEITOZA.

Giant Brazilian Seahorse *Hippocampus* cf *reidi*
Undetermined species

Uncertain distribution, probably restricted to eastern Brazil. Shallow, protected coastal bays and estuaries. Has low coronet and moderately large tubercles on its trunk. Grows large, height to about 26 cm.

A

B

H. cf reidi. Brazil. **A** female. Ilha Grande Bay. Ricardo Z. P. GUIMARAES. **B** male. Todos os Santos Bay. Depth 6 m. Bertran M. FEITOZA.

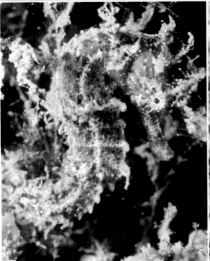

C

D

E

H. cf reidi. **C–E** females, colour variations. Nacau, Brazil. Depth 1 m. Bertran M. Feitosa.

Pacific Seahorse
Hippocampus ingens

Hippocampus ingens Girard, 1858.
San Diego, California
Hippocampus ecuadorensis Fowler, 1922.
Bahia, Ecuador.

East Pacific, from California to Peru. Closely related to West Atlantic species. Highly variable in colour, but adults usually with close-set, white scribbly thin lines running along head and body. Some slight differences in colour and shape of coronet between northern and southern populations may warrant further investigation. Mainly found on reefs, clinging to sponges or branches of corals. Depth range about 6–25 m but may venture deeper. Also has been found in surface waters on floating weeds. Height to 26 cm.

H. ingens. Male. Galapagos Islands. Height 20 cm. Jerry ALLEN.

H. ingens. Galapagos Islands. Small male about 14 cm height. Frank Schneidewind.

H. ingens. Close-up of **A.**

H. ingens. **D** female. Galapagos I. Frank Schneidewind. **E & F** female. Sea of Cortez. Howard HALL.

Long-snout Seahorse
Hippocampus reidi

Hippocampus reidi Ginsburg, 1933.
Granada, West Indies.

West Atlantic, widespread from Florida to West Indies. A similar species *H. deanei* Duméril, 1857, Sierra Leone, off West Africa. Extremely variable in colour, matching various sponge colours. Both sexes with pale saddle-like banding. Various soft-bottom habitats, usually clinging to slender sponges or gorgonians with matching colours. Occurs in depths to at least 50 m. Height to 15 cm.

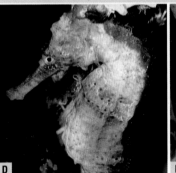

A B

H. reidi. Female yellow, male dark red. Bonaire, Caribbean. Roger C. STEENE.

C

H. reidi. Male. Paul HUMANN.

D E

H. reidi. Female variation. St. Vincent, Caribbean. Roger C. STEENE.

F

H. reidi. Male, variation. Paul HUMANN.

G

H. reidi. Female, variation. Michelle HALL.

H

H. reidi. Female. St. Vincent, Caribbean. Roger C. STEENE.

A

H. histrix. Osezaki, Japan. Height about 5 cm. Hiroyuki UCHIYAMA.

Thorny Seahorse *Hippocampus histrix*

Hippocampus histrix Kaup. 1856. Japan.

West Pacific, mainly known from Japan to Indonesia and Coral Sea. Rare in northern Australia where several other spiny species occur. Reports from elsewhere need to be verified as most spiny species were going under the name *H. histrix*. Primarily lives at moderate depths of about 15 m or deeper, on soft bottom with soft corals and sponges, but occasionally found inhabiting shallower algae-rubble reef zones in about 10 m depth. Highly variable in colour from bright yellow to red or green to match surroundings. Red colours are camouflage in deep water where the colour is neutralised to grey. A distinct species with its spiny appearance and long snout, latter appears very long in juveniles and may elongate in large males. The sparse white barring on the snout, as shown in the photographs, is diagnostic. Height to about 15 cm.

Because of their moderate depth preference they are not often collected for the aquarium trade. If they are offered for sale, this species is probably much more demanding, needing top-quality conditions such as in an aquarium that supports soft corals and has low light conditions. Like any syngnathid, anemones or stinging corals should not be sharing their captive home.

Spelling *hystrix,* found in some literature, is based on another description of the same species that is a junior synonym.

B

H. histrix. Osezaki, Japan. Height about 7 cm. Hiroyuki UCHIYAMA.

C

D

E

H. histrix. **C** female, Mabul, Malaysia. **D** female, **E** male, adults on soft bottom. Lembeh Strait, Sulawesi, Indonesia. Roger C. STEENE.

Spiny Seahorse *Hippocampus jayakari*

Hippocampus jayakari Boulenger, 1900. Muscat.

Red Sea and Arabian Sea. Often confused with *Hippocampus histrix* and Indian Ocean records need to be investigated. A species reported from south Africa as *H. histrix* is probable another, perhaps undescribed taxon. *H. jayakari* has been found in rubble-algae habitats with sparge seagrasses and soft-bottom habitat on sponges to about 20 m depth. Hatchlings apparently settle on the substrate, which may account for a limited geographical distribution. Height to 14 cm.

Remarks: This species is kept in the Eilat public aquarium and apparently is easy to maintain.

A

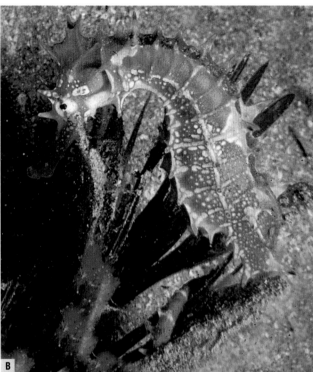
B

C

H. jayakari. Saudi Arabia, Red Sea. **A** male. **B** female. Helmut DEBELIUS.

H. jayakari. Red Sea. Male. Thomas PAULUS.

D

H. jayakari. Eilat, Red Sea. Brooding male, height about 14 cm. Frank SCHNEIDEWIND.

A

H. elongatus. Woodman's Point, Fremantle, Western Australia.

Hippocampus elongatus Castelnau, 1873.
Fremantle, Western Australia.
Hippocampus subelongatus Castelnau, 1873.
Fremantle, Western Australia.

Mainly sub-tropical Western Australia. Doubtfully recorded from South Australia. However, much of the area in southern Western Australia is unexplored and this species may occur further south than presently known. Large numbers of *Hippocampus elongatus* congregate in the lower reaches of the Swan River in early summer when many of the crustaceans are spawning and provide plenty of food for their offspring. Few can be found there during winter when the rain reduces the salinity. It occurs commonly in sheltered coastal bays in mixed reef and vegetation habitats to about 10 m depth, but may move to deeper water in non-breeding periods. Height to 20 cm.

Remarks: this species is collected in the wild for the aquarium trade. The fluctuation of numbers in the Swan River system was blamed on collectors, but is really due to seasonal conditions. The number of specimens collected for aquarium purposes is minuscule compared to the wild population. Closely related to the more tropical *H. angustus* which has longer snout and sharp spines on head and body.

This species is easily maintained in aquariums and exported in low numbers (about 250 per year).

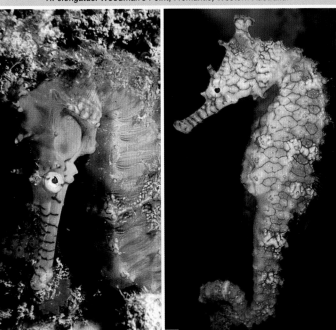

B

H. elongatus. Woodman's Point, Fremantle, Western Australia.

C **D** **E**

H. elongatus. **C** Woodman's Point, Fremantle, W. A. **D** & **E** female & male respectively. Perth, W.A. Mark NORMAN.

Eastern Spiny Seahorse *Hippocampus hendriki*

Hippocampus hendriki Kuiter, 2001. Queensland.

North-eastern Queensland to Keppel Island. This species was reported as *Hippocampus histrix*, *H. angustus* or *H. spinosissimus*. *H. hendriki* is distinguished from those in spine arrangements and colour. It differs from the sympatric *H. queenslandicus* in longer spines and having a nose-spine. *H. hendriki* is known from trawled material, relatively shallow to about 50 m from open substrate adjacent to inner reefs. Height to 12 cm.

A **B**

H. hendriki. **A** East off Newcastle Bay, Qld. D. 20 m. Female. H. 85 mm. **B** W. off Thursday Island. D. 12 m. H. 7 cm. CSIRO FISHERIES.

Big-head Seahorse
Hippocampus grandiceps

Hippocampus grandiceps Kuiter, 2001. Gulf of Carpentaria, Queensland.

Appears to be restricted to the eastern part of the Gulf of Carpentaria. Its limited distribution may reflect a unique habitat. The area lacks the coral reef formations that are prominent habitat features in the western part of the Gulf. It is one of the spiny species, but has reduced spines and the head is held close against the chest, indications that it probably lives in weedy habitats. All the specimens held in museums were collected by dredging in depths of 10–12 m. It is closely related to *Hippocampus angustus* but is a smaller, stockier species with a proportionally larger head, its head about equal to trunk in lengths. Largest specimen measured is 11 cm in height.

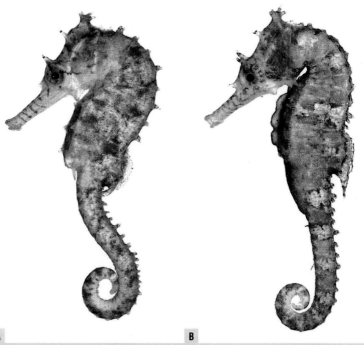

A **B**

H. grandiceps. Paratypes, Gulf of Carpentaria, Qld. D. 10 m. **A** female. H. 69 mm. **B** male. H. 75 mm

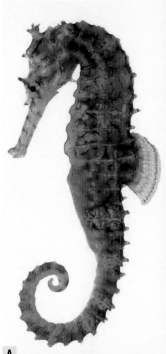

A

H. angustus. Male. **E** Monte Bello Is, W.A. Height 14 cm. Barry HUTCHINS.

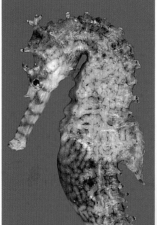

B　　　　　　　CSIRO FISHERIES

H. angustus. NW of Port Hedland, WA. D. 63 m. Male, height 11 cm.

Western Spiny Seahorse *Hippocampus angustus*

Hippocampus angustus Günther, 1870. Shark Bay, W.A.

Western Australia, Shark Bay to Dampier Archipelago. Has high coronet with 5 spines and sharp spines on almost every ridge on the back, above and in front of eyes. Females are more spiny than males. Snout long, usually banded. Ventral trunk ridge with scalloped edge in male, and low downward spines in females. Algal reef, from 12 to about 25 m depth, occasionally trawled deeper. Height to at least 16 cm.

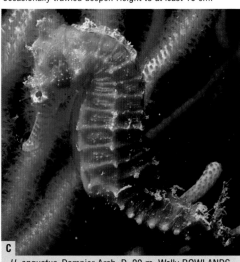

C

H. angustus. Dampier Arch. D. 20 m. Wally ROWLANDS.

Northern Spiny Seahorse *Hippocampus multispinus*

Hippocampus multispinus Kuiter, 2001. Arafura Sea.

Known from northern Australia and southern Papua New Guinea. Males much less spiny than females, especially over the back of the trunk. Usually deep water. Most Australian material was trawled in depths over 40 m but few specimens came from shallow water in Darwin Harbour. Height to 15 cm.

Was confused with *Hippocampus histrix* and *Hippocampus spinosissimus* Weber, 1913, a small species from 70 m depth near Komodo that is only known from the types and some photographs (see next page).

A

H. multispinus. Milne Bay, PNG. Bob Halstead.

B

H. multispinus. Aquarium. From Darwin Harbour. Helen LARSON.

34

Zebra-snout Seahorse *Hippocampus barbouri*

Hippocampus barbouri Jordan & Richardson, 1908. Philippines.

Known from Philippines and northern Indonesia, but may occur in adjacent waters of Japan. Previously confused with other spiny species, e.g. *H. histrix, H. angustus* and *H. spinosissimus*. Recognised by the spiny coronet, striped snout, double cheek-spines and thick spine in front of eyes on the snout. Unlike most seahorses, this species is often found clinging to hard corals and doesn't seem to be worried about stingers. Known depth range 6–10 m. Height to 15 cm.

A

B

H. barbouri. Philippines. **A** female, height ~9 cm, aquarium. **B** male, height ~14 cm. Roger C. STEENE.

Hedgehog Seahorse
Hippocampus spinosissimus

Hippocampus spinosissimus Weber, 1913. Sapeh Strait.

Only known from the 2 types trawled from 70 m depth near Komodo and photographs that appear to be of this species. Habitat was described as sand and scallops. Illustration by Weber (**A**), is clearly a fully developed male. Meristics are: dorsal fin 17; pectoral fin 15; trunk rings 11 and tail rings 34. Total length was given as 70 mm. Specimen in Philippines was observed for 1 month on the lower part of drop-off with some sponges in an area influenced by strong tidal currents. In that time it accumulated some algae-growth and the specimen from Sulawesi below is thought to be this species as well, and not a juvenile *H. molluccensis* as first thought. It seems that this is a deep water species that occasionally may be brought up by strong currents to the shallower depths.

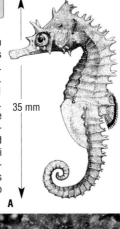

35 mm

A

B

C

H. spinosissimus. **B** northern Sulawesi, Indonesia. H. 4 cm. Roger STEENE. **C** southern Cebu, Philippines. D. 15 m. H. 5 cm. Kazunori (Garuda) IGARASHI.

Half-spined Seahorse
Hippocampus semispinosus

Hippocampus semispinosus Kuiter, 2001.
East Alas Strait, Indonesia.

Known from southern Indonesia, Bali to Timor Seas, and possibly northwestern Australia. Males usually dull yellowish brown with a series of distinctive white blotches on the trunk. Females cream to yellow or reddish brown with broad saddle like banding behind the head and near dorsal fin. Snout elongates in large adults and this species is distinguished from other similar sub-smooth species by the more slender trunk. It often has algae growth on the spiny ridges. Lives on open substrates, from near mangroves to deeper muddy channels, or deep offshore where sometimes trawled. Height to 14 cm.

H. semispinosus. Juvenile, height 7 cm. Gilimanuk, Bali, Indonesia. Miki TONOZUKA.

H. semispinosus. Gilimanuk, Bali, Indonesia. **B** male, height 12 cm. Takamasa TONOZUKA. **C** juvenile, height 8 cm. Alex STEFFE.

Fourstar Seahorse *Hippocampus arnei*

Hippocampus arnei Roule 1916. Mekong R., Laos and Thailand border.

Must have originated from the sea. Of the two lectotypes, only one is this species, the other is actually *H. barbouri*. First published as *H. aimei*, but corrected soon after to *H. arnei*, as he named it after the Arne brothers who donated the specimens. Occurs in China Seas and Philippines. Inshore habitats, algae reef. Like the '*kuda*' group, it has the upper spine on the shoulder ring (in front of the pectoral fin base) near the gill opening, that is near the pectoral fin in the other spiny species. It usually lacks a spine in front of the eyes (nose-spine) or it is very small. The main lower shoulder ring spines point outward and a smaller secondary one forwards. It has a small but spiny coronet that often has 4 instead of the usual 5 spines in similar species. Height to 10 cm.

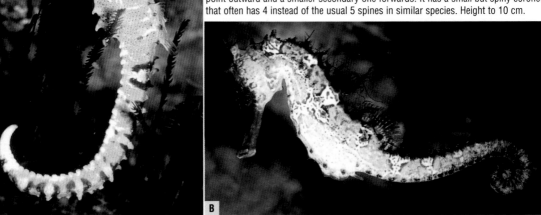

H. arnei. Philippine specimens, aquarium. **A** THE SEAHORSE TRUST. **B** Frank SCHNEIDEWIND.

Moluccan Seahorse
Hippocampus moluccensis

Hippocampus moluccensis Bleeker, 1852.
Ambon, Indonesia.

A common species in Ambon and eastern Sulawesi, possibly restricted to Moluccan Sea. Similar to *Hippocampus taeniopterus*, but coronet is usually more distinct. Snout thick and its length equal to distance from eye to gill-opening. Males drab to ornamental red and purplish when in soft corals, usually a white saddle on neck followed by 3 more until dorsal fin, and white striations on cheek. Open sandy and muddy substrates from shallow protected bays on soft-bottom slopes to at least 20 m depth. Occasionally found in pairs. Height to 16 cm.

Remarks: Bleeker's types are housed in Museum Victoria, Melbourne, and matches this species. He obtain 3 more specimens in Ambon at the same time, 8–11 cm, which he named *Hippocampus taeniopterus* (see P. 45) and noted that these differed from *H. moluccensis* in snout length and in having a black submarginal stripe in the dorsal fin.

A
H. moluccensis. Lembeh Strait, Sulawesi, Indonesia. Male, height 15 cm.

B
H. moluccensis. Lembeh Strait, Sulawesi, Indonesia. Female, height 16 cm. Scott MICHAEL.

D
H. moluccensis. Lembeh Strait, Sulawesi, Indonesia. Male, height 15 cm. Janine MICHAEL.

C
H. moluccensis. Lembeh Strait, Sulawesi, Indonesia. Male, height 15 cm. Takamasa TONOZUKA.

Collared Seahorse *Hippocampus jugumus*

Hippocampus jugumus Kuiter, 2001. Lord Howe Island.

Known from a single specimen that was collected as far back as 1925. No further details are known. The specimens appears to have been dried, perhaps found on beach, prior to its preservation. It is unusual in having 12 trunk rings and the shoulder rings being confluent over the neck-ridge, forming a continuous collar. It is very spiny and many on the head are double spines. There are no similar species known in the area. Further meristics are: dorsal fin 20; pectoral fin 16; tail rings 37; dorsal fin base over 5 rings. Height is 44 mm.

H. jugumus. Type specimen from Lord Howe Island.

37

Wing-spined Seahorse *Hippocampus alatus*

Hippocampus alatus Kuiter, 2001. Cape York, Queensland.

Known from Gulf of Carpentaria, northern Australia, and southern Papua New Guinea. Juveniles with several pairs of flat and outward pointing spines over the back that reduce in size with growth. Often with filaments on the larger spines on the head and back. Usually found on remote outcrops of debris or corals that provide shelter and a hold. Found in deep current prone channels between reefs or islands, in depths over 20 m. Height to 12 cm.

A   B

H. alatus. Milne Bay, PNG. Females about 15 cm in height at 25 m. Bob HALSTEAD.

Queensland Seahorse *Hippocampus queenslandicus*

Hippocampus queenslandicus Horne, 2001. Cape York, Queensland.

Only known for certain from Southport, Queensland and north to Papua New Guinea. Replaced by the similar and larger *H. tristis* further south. Most specimens are known from trawls near reefs to 63 m depth and rarely seen in less than 20 m depth. Deep water specimens are usually red or orange, colours that are probably similar to corals and sponges living at that depth. Height to 13 cm.

Remarks: Spiny young-adults have been identified as *H. spinosissimus,* and less spiny large adults as *H. kuda.*

A

H. queenslandicus. Female. Off Townsville, from ~40 m depth.

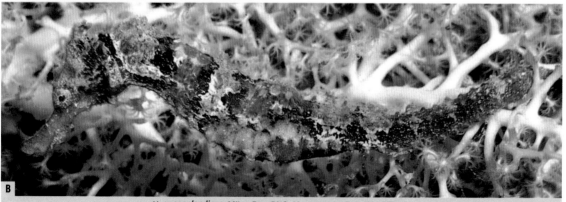

B

H. queenslandicus. Milne Bay, PNG. Young male. Bob HALSTEAD.

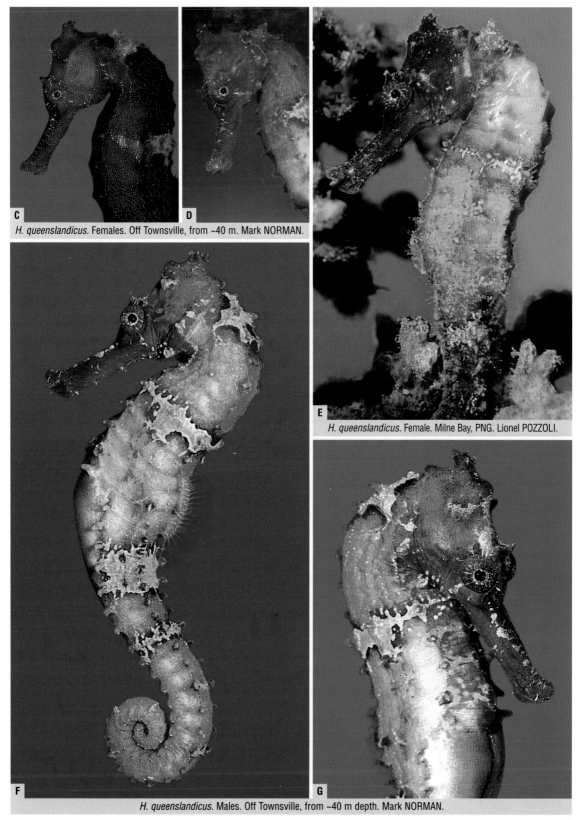

H. queenslandicus. Females. Off Townsville, from ~40 m. Mark NORMAN.

H. queenslandicus. Female. Milne Bay, PNG. Lionel POZZOLI.

H. queenslandicus. Males. Off Townsville, from ~40 m depth. Mark NORMAN.

Flores Seahorse *Hippocampus polytaenia*

Hippocampus polytaenia Bleeker, 1852.
Flores, Indonesia.

Appears to be restricted to Flores Seas, but probably ranges to the Moluccen region. At some stage it was thought to be a juvenile of *H. kuda* but it is more spiny and different in shape, also *H. kuda* was not found in Maumere Bay during many surveys over about 10 years and probably doesn't occur there. This fits the distribution pattern of many other Indonesian fishes. *H. polytaenia* is mainly found inshore, on shallow reef flats in sargassum weeds and in silty habitats, sometimes in open substrates with sparse vegetation. Yellow with pinkish banding when in weeds and to drab grey or black on open muddy substrates. To about 5 m depth. Probably a small species, known height to 9 cm.

A Bleeker's illustration of the type.

B *H. polytaenia*. Flores, Indonesia. Height 8 cm.

C *H. polytaenia*. Maumere, Flores, Indonesia. Height 8 cm.

Tiger-tail Seahorse *Hippocampus comes*

Hippocampus comes Cantor, 1850.
Pinang, Malaysia.

Andaman Sea to Bintan, near Singapore, primarily Malacca Strait and adjacent waters. Possibly ranges into China Seas to Philippines or east to Java. A distinctive species in colour in which large adults are often bright yellow. Males often black with yellow blotches over the back and bands on the tail, especially showing towards the tip. It has double cheek spines that are clearly visible in most pictures included here. This species is mainly found in pairs on reefs in rich soft coral areas and usually in depths over 20 m. They are usually seen at night as they sleep higher above the substrate for safety. Height to 16 cm.

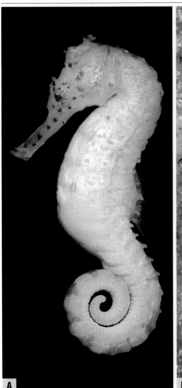

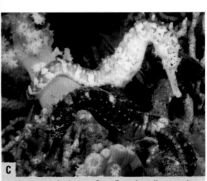

A *H. comes*. Bintan, Indonesia. Yellow Male. Clay BRYCE.

B *H. comes*. Andaman Sea. Female. Mark STRICKLAND.

C *H. comes*. Andaman Sea. Female yellow, male black. Mark STRICKLAND.

Lemur-tail Seahorse *Hippocampus mohnikei*

Hippocampus mohnikei Bleeker, 1851.
Kaminoseki I, Japan.

Izu region, to China Seas, and the Philippines (if the same). Seagrasses and soft-bottom, in 1–30 m depth. Has double cheek spines, a small nose-spine, and distinct coronet in young. Closely related to *Hippocampus comes,* usually called this in the Philippines, but called *H. kuda* or *H. kelloggi* in Japan. Reddish brown to black with or without white saddles or bands. Tail long and is variably marked with narrow to distinctive white bands, sometimes yellowish at the tip. Height to 16 cm.

Remarks: In his *Hippocampus moluccensis* description, Bleeker remarked that he distinguished *H. mohnikei* from other seahorses in his collection by the diagnostic white banding on the tail. No other species in Japan has such bands.

A

H. mohnikei. Kashiwajima, Japan. Depth 22 m. Female, height 10 cm.

B *H. mohnikei.* Juveniles raised in Aquarium.

C *H. mohnikei.* Kashiwajima, Japan. Height 10 cm.

D *H. mohnikei.* Aquarium. H. 10 cm.

E
F

H. mohnikei. Females, collected from 1–2 m, in southern Miyazaki, Japan. Hiroyuki TANAKA.

G *H. mohnikei.* Male, aq., Japan. H.16 cm.

Reunion Seahorse *Hippocampus borboriensis*

Hippocampus borboriensis Duméril, 1870. Réunion Is.

West Indian Ocean, previously confused with *H. kuda* from the West Pacific, or *H. whitei*, an Australian endemic. Shallow protected bays to deep water where found in soft-bottom and sponge habitat. Like other seahorses in deep water, they can be very colourful and usually lack filamentous appendages that the same species often has in shallower depths. Usually brownish grey with short algae growth on the skin that hides some of the colour in the shallows. In deep water they lack any algae growth vary from almost pure white to reddish brown and may have a distinctive banded pattern. Known depth range 5 to 60 m. Height to 9 cm.

A **B**

H. borboriensis. **A** male 60 mm & **B** female, 45 mm.
Seychelles, from 57 m depth. John E. RANDALL

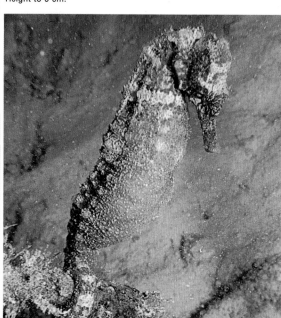

C **D**

H. borboriensis. Seychelles. Depth 5 m. Height 9 cm. Neville COLEMAN

Hong Kong Seahorse *Hippocampus* cf *kuda*
Undetermined species

A smooth species from the Hong Kong region in coastal waters. Snout very thick, lacks any nose spine or bump, and has a single cheek spine. Looks similar to *Hippocampus kuda*, but this has a slightly longer snout and small nose-spine. Height to ~20 cm.

A **B**

H. cf *kuda.* Hong Kong. Depth 8 m. Height about 15 cm. Rowena YUE.

Drab Seahorse *Hippocampus fuscus*

Hippocampus fuscus Rüppell, 1838.
Jeddah, Red Sea.

Little known species from the Red Sea and Arabian seas. Probably was confused with *H. kuda* before. Pale yellowish brown to dark-grey to black, usually with large pale areas over the back. A large whitish area following the head maybe diagnostic for this species and can be distinguished in some of the paler individuals. Shallow protected waters on the edges of algal-reefs or seagrass beds in 1–10 m depth. Grows moderately large, known height to 15 cm.

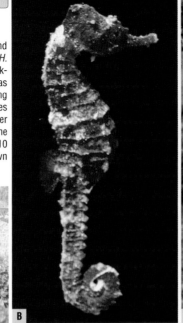

H. fuscus. Aqaba, Jordan. All photographs Thomas PAULUS.

Sad Seahorse *Hippocampus tristis*

Hippocampus tristis Castelnau, 1872.
Melbourne fish market, Australia.

Know from trawled and beached specimens off northern New South Wales, Lord Howe Island and southern Queensland. It has been identified as *H. whitei, H. kuda & H. kelloggi*, the name usually in relation to size. Juveniles are spiny and large adults smooth. It is distinguished from most other similar species by the high fin counts in both dorsal and pectoral fins which is typically 18 or 19, and the double blunt spines ventrally on the shoulder ring, below the pectoral fins. Reaches a height of 23 cm.

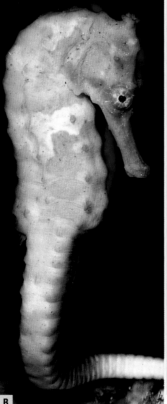

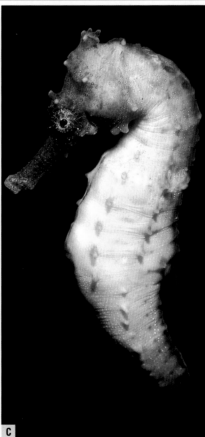

H. tristis. Aquarium, from off northern NSW. Heights: **A** juvenile 5 cm, **B** female 15 cm, **C** male 20 cm.

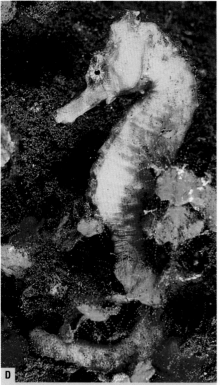

Hippocampus kuda Bleeker, 1852. Singapore.

Maldives, Sri Lanka, Andaman Sea, Singapore and western Indonesia to Ryukyus, Japan. Drab brown to black, often with long filaments on crown and snout. Sub-marginal black line in dorsal fin. Square head-profile over eye. Males usually drab with numerous small dark spots over the body and females often all black, sometimes plain yellow. Snout short in young, elongating with growth. Occurs in estuaries, harbours, and lower reaches of rivers, entering brackish water. To 15 cm height.

Remarks: Most records of *H. kuda* are based on various smooth species, especially the next one, *H. taeniopterus*.

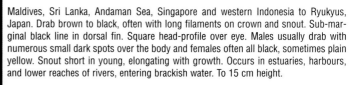

A

B

C

H. kuda. **A & B** juveniles, aquarium imports, probably from Sri Lanka.

H. kuda. From Keppel Harbour, Singapore. Underwater World.

D

H. kuda. Yellow female, height 15 cm, at 15 m depth. Tulamben, Bali, Indonesia. Takamasa TONOZUKA.

E

H. kuda. From Bleeker's Atlas.

F

H. kuda. Male. Aquarium import.

Common Seahorse *Hippocampus taeniopterus*

Hippocampus taeniopterus Bleeker, 1852. Ambon.
Hippocampus melanospilos Bleeker, 1854. Ambon.

Moluccen Seas to Sulawesi and Bali, and to north and north-eastern Australia. Females are sometimes yellow with several large dark spots on the trunk (*H. melanospilos* form). Males usually drab with striations over the head and small black spots over the trunk. Usually found along margins of seagrass beds, rarely deeper than 15 m. Adults occur usually in pairs. To 22 cm height.

Remarks: This species is generally referred to as *H. kuda*.

H. taeniopterus. Juvenile, height 40 mm. Flores, Indonesia. Depth 3 m.

H. taeniopterus. South of Bitung, northern Sulawesi. Pair, about 18 cm in height.

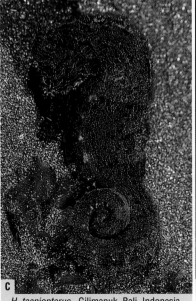

H. taeniopterus. Gilimanuk, Bali, Indonesia.

H. taeniopterus. Milne Bay, PNG. **C** Female. Bob HALSTEAD. **D** male, height 18 cm.

Egyptian Seahorse
Hippocampus suezensis

Hippocampus suezensis Duncker, 1940.
Red Sea.

Probably restricted to the Red Sea and Arabian Seas. Previously reported as *H. kuda* from Arabian Seas but is closely related to the Japanese *H. kelloggi,* sharing the slender trunk, and additional similar species maybe elsewhere. Dusky brown to black with tiny yellow spots, often forming series of striations. Female sometimes pale yellow or cream with dark banding, the dark parts with series of tiny yellow spots. Lives at moderate depths in soft corals. In Oman all specimens were photographed in about 30 m depth. Height to at least 22 cm.

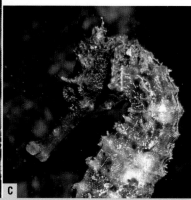

H. suezensis. Large female, height 22 cm. Muscat, Oman. Phil WOODHEAD.

H. suezensis. **D** male, height 18 cm, **E** & **F** female, height 14 cm Oman. Phil WOODHEAD.

Great Seahorse *Hippocampus kelloggi*

Hippocampus kelloggi Jordan & Snyder, 1902. Kyushu, Japan.

Tropical to sub-tropical southern Japan, probably ranging into China Seas. Reported as *H. kuda*, but much more slender and closely related to *H. suezensis.* Trunk long, and tip of snout expanded. Adults with distinctive fine white lines and spots in dark parts of head and trunk (typically as in **D)**, that may persist in dried specimens. Brownish to black, but juveniles and female sometimes cream or yellow, often with light saddles or patches. Small juvenile (**A**), less the 10 cm height, with very slender trunk and series of small white spots that appear in the dark areas on the head first. Mainly found on soft-bottom in depths over 20 m. Height to at least 25 cm (several specimens were measured at this size at Kashiwajima, Japan, where it is common), probably reaches 28 cm.

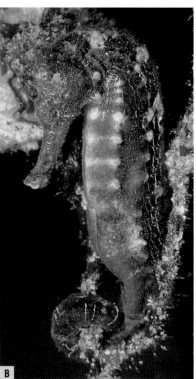

H. kelloggi. Juv. Hachijoh-jima, Japan. Shoichi KATO.

H. kelloggi. Kashiwajima, Japan. Male, height 24 cm.

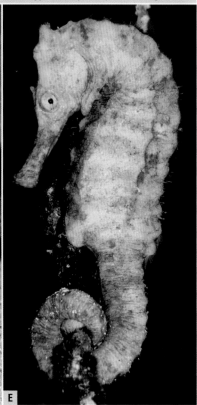

H. kelloggi. Kashiwajima, Japan. Depth 25 m. Height 24 cm. **C** Male, **D** Female.

H. kelloggi. Kochi, Japan. Female, Tomonori HIRATA.

47

Camel Seahorse
Hippocampus camelopardalis

Hippocampus camelopardalis
Bianconi, 1855. Mozambique.

Southern East African coast. Estuarine species, mainly found in seagrass and algae beds or shallow reef. Dusky to marbled with obvious eye-like spots serve to create a crab-like face when it bends over for protection from predators when in seagrasses. This pattern is also in an Australian that lives in similar habitat (P.50). A small species, to about 9 cm in height.

Remarks: Records of *H. fuscus* from South Africa are probably based on this species.

H. camelopardalis. **A** Natal, Afr. Dennis KING. **B** Inhaca, Mozambique. Phil HEEMSTRA.

Krysna Seahorse *Hippocampus capensis*

Hippocampus capensis Boulenger, 1900. S. Africa.

Southern coast of South Africa. Mainly on algal reefs in estuaries about 20 m depth. Populations can fluctuate greatly in shallow estuaries and reported kills from washed up specimens can be caused by large amounts of freshwater input. Height to 9 cm.

H. capensis. Male. Krysna, South Africa. Jan C. POST.

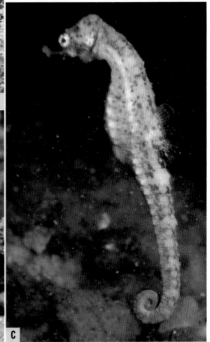

H. capensis. Krysna, South Africa. **B** female. **C** young male. Guido ZSILAVECZ.

West African Seahorse
Hippocampus deanei

Hippocampus deanei Duméril, 1857.
Sierra Leone.

Only known from the tropical West African coast and one of the least known species, due to the lack of diving facilities there. Found in protected coastal bays and harbours, where it lives on open muddy substrates to about 25 m depth. Height to 20 cm.

Synonymy

H. punctulatus Kaup, 1856 is also this species, but name unavailable, preoccupied by *H. punctulatus* Guichenot, 1853 (= *H. erectus* Perry, 1810).

H. deanei. Depth 24 m. Height 20 cm. São Tomé, on equator, off West Africa. Peter WIRTZ.

Smooth Seahorse
Hippocampus kampylotrachelos

Hippocampus kampylotrachelos Bleeker, 1854. Priaman, Sumatra, Indonesia.

Known from the southern Indonesian main islands from Sumatra to west Bali, possibly Sulawesi (*H. manadensis*), and one specimen was found amongst birds nesting on Ashmore reef in the Timor Sea. Slender species with a low crown formation on top of the head, and upward directed nape-spine. Dusky brown with fine pale spotting over the head and back, and dorsal fin with thin dark stripes, one along centre and another near its margin. Numerous lines radiating from pupil in the eye. Shallow coastal muddy estuaries and probably deep offshore. Height to at least 22 cm.

H. kampylotrachelos. Depth 4 m. Height 22 cm. Gilimanuk, Bali, Indonesia. Akira OGAWA.

Hawaiian Seahorse
Hippocampus fisheri

Hippocampus fisheri
Jordan & Evermann, 1903. Hawaiian Is.

Probably restricted to the Hawaiian region and where it is probably the only species there. Reports of *Hippocampus kuda* and *H. histrix* are probably based on adult and juvenile respectively of this species. Little known species that is usually found drifting in open waters near the surface and one specimens reported as *H. histrix* from 55 m trawl. Juvenile with knobby blunt spines and adults more smooth. Height to about 8 cm.

H. fisheri. Surface waters, night. Male, height ~6 cm. Hawaii. Chris NEWBERT.

False-eyed Seahorse *Hippocampus biocellatus*

Hippocampus biocellatus Kuiter, 2001. Shark Bay, WA.

New species, only known from Shark Bay to Dampier, WA. Recognised by the eye-marks on the back in both sexes, facing laterally that when in the weed may be used to bluff potential predators. Snout rather short and trunk deepens with age. Males develop deep keel of soft skin along median ridge of trunk and adults with large tubercles along back ridges. Lives in algae reefs and seagrasses in protected shallow bays to a few meters depth. Height to about 11 cm.

Flat-face Seahorse *Hippocampus planifrons*

Hippocampus planifrons Peters, 1877. Shark Bay, WA.

Shark Bay to Exmouth, WA. Snout rather short when young, elongating in adults. Trunk slender, including in males. Ridges with small tubercles, becoming smooth with age. Snout spotted. Lives in algae and rubble reefs in shallow bays to 20 m. depth. Height to about 12 cm.

A *H. biocellatus*. Female. Shark Bay, W.A. Height 65 mm. Barry HUTCHINS.

B *H. biocellatus*. Dampier, W.A. Depth 5 m. H. 90 mm. Wally ROWLANDS.

H. planifrons. Shark Bay, W.A. Height 53 mm. Barry HUTCHINS.

Low-crown Seahorse *Hippocampus dahli*

Hippocampus dahli Ogilby, 1908. Noosa, Qld.

Coastal Queensland to Darwin, N.T. and known from a photograph taken of the zebra-form in Milne Bay, PNG. Males usually from drab grey-brown with yellowish chest or pouch to near black. Females from drab grey-brown to cream with dark scribbles, sometimes with zebra-pattern (**A**). Northern variation (**C**) shows series of pale spots on head and across trunk. Four spines along base of dorsal fin and a lateral line present with pores to about 20th tail ring. In channels of estuaries and on soft bottom offshore. Height to about 14 cm.

A *H. dahli*. **A** zebra-form, female. Height 12 cm. Moreton Bay, Qld. **B** female. Qld, Australia. Ern GRANT. **C** male. Darwin, NT, Australia.

Zebra Seahorse
Hippocampus zebra

Hippocampus zebra Whitley, 1964.
Swain Reefs, Qld, Australia.

Only known from a few specimens off the eastern Australia coast from Cape York to northern NSW. All known specimens of *H. zebra* have the distinctive zebra-like striped pattern, but two other species can have this on occasion as well: the closely related *H. montebelloensis* and the low crowned *H. dahli*. Latter has a very low coronet, different meristic counts, the most obvious the 4 or 5 tubercles along dorsal fin base versus only 3 in *H. zebra*. A zebra-striped *H. montebelloensis* can be most similar but is generally more spiny at similar sizes, especially when small, and has a larger head. Known depth range about 20–60 m depth. Soft bottom habitat, probably on black-coral fans or gorgonians where it may mimic the black and white basket stars. Height to about 9 cm.

A

H. zebra. Type, female. Swain Reef, Qld. Anthony Healy, Australian Museum.

B

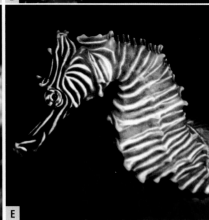

C

H. zebra. Male, height 71 mm. Trawled from off Cape Grenville, northern Qld, Australia.

D **E**

H. zebra. Female specimen collected by trawl from off Tweed Heads, NSW, Australia in 60 m depth. Height 9 cm. **B** Tom Bowling. **D** & **E** Neville Coleman.

Monte Bello Seahorse
Hippocampus montebelloensis

Hippocampus montebelloensis Kuiter, 2001.
Monte bello Islands, WA.

Known from 4 specimens from Exmouth Gulf to the Dampier Archipelago, WA. The holotype was collected on the surface at night at the Monte Bello Islands, clinging to lose *Sargassum* with the bottom 5 m below. Other specimens were collected by trawl in depths between 15 and 35 m. This species is closely related to *Hippocampus zebra* and has similar meristics and morphology. One specimens has a very similar striped pattern, one has a fine-lined pattern that also occurs in *H. dahli* and the other two specimens are plain. Morphology differs primarily in *H. montebelloensis* being more spiny and having a larger head in relation to the trunk (about 74% in *H. zebra* versus 95% in *H. montebelloensis*. All specimens collected are female, largest is 78 mm in height.

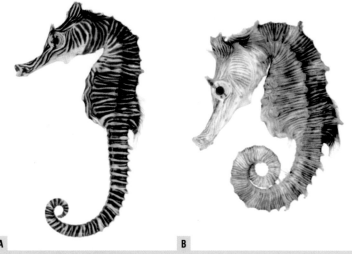

A **B**

H. montebelloensis. Females. Heights: **A** 70 mm, **B** 65 mm. Dampier, WA.

Japanese Three-spot Seahorse
Hippocampus takakurai

?Hippocampus trimaculatus Leach, 1814.
As Indian & China Seas, but holotype probably China Seas.
Hippocampus takakurai Tanaka 1916.
Chiba Pref, Japan.

Chiba Prefecture, Japan, and ranging south to sub-tropical China Seas if representing a single species. Large and pale male are distinctive with dark spots along upper trunk ridges (**B**), but general colours can vary from grey to dark brown or near black. Large adults usually with some yellow spotting on the head and trunk. Snout about 1/2 head-length. Coronet distinctive in young. A recurving single spine below the cheek and 3 spines along dorsal fin base. Lives at moderate depths on algal reef and soft-bottom substrates to at least 30 m. Height to 16 cm.

Possibly a synonym of *H. trimaculatus,* that looks very similar.

H. takakurai. Young male, height 10 cm. IOP, Izu Pen. Japan. Depth. 10 m. Hajime MASUDA.

H. takakurai. Male, 15 cm. Depth 20 m. Southern Japan. Yusuke YOSHINO.

H. takakurai. Female, height 12 cm. Osezaki. Japan. Hiroyuki YAMAZAKI.

Dwarf Seahorse *Hippocampus zosterae*

Hippocampus zosterae Jordan & Gilbert, 1882. Florida.

West Atlantic, Florida, Bahamas, Bermuda and Gulf of Mexico. Shallow estuarine species found in seagrasses, usually in depths of a few meters. Occurs in pairs and small groups. Male may take brood twice a month. Brood comprises few large eggs, usually 10–50, depending on size of male, and incubation period is only about 10 days. Young grow fast and this species can have 3 generations a year. A tiny species, height usually less than 30 mm.

An easy species to keep in small aquarium, tolerant to various conditions and can be kept in small groups. Lives for 1–2 years in the wild, but may last longer in captivity.

A B

H. zosterae. Aquarium. Scott MICHAEL.

C

H. zosterae. Gravid female. Aquarium. Scott MICHAEL.

D

H. zosterae. Male. Aquarium. Scott MICHAEL.

Bullneck Seahorse
Hippocampus minotaur

Hippocampus minotaur Gomon, 1997. South-Eastern Australia.

Only known from southern New South Wales to the Bass Strait region. All specimens came from trawls deeper than 60 m. All specimens are rather smooth and lack spines, suggesting that they may live on the branching sponges, or perhaps even inside the tubular parts. Life colours are unknown and preserved specimens brownish or cream with brown spots with cream centres. A small species, height to about 45 mm.

Remarks: A report of this species from much shallower surveys off Bass Point, south of Wollongong, NSW, is probably incorrect. The specimens were lost and it seems now more likely that it was based on the pygmy pipehorse *Idiotropiscis lumnitzeri* that seahorse-like.

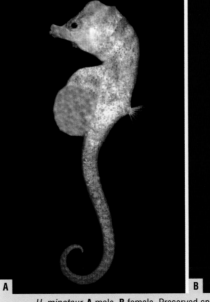

A B

H. minotaur. **A** male, **B** female. Preserved specimens, computer modified.

A

H. japonicus. Izu, Japan. Brooding male, height 10 cm. Yusuke YOSHINO.

Mossy Seahorse _Hippocampus japonicus_

Hippocampus japonicus Kaup, 1856. Nagasaki, Japan.

A Japanese sub-tropical species from the southern Japanese Seas, from the Kyushu region to western Honshu and Izu Peninsula. A second form or species (**E–G**) along Honshu's north coast, known as 'Sangotatsu' or Northern Seahorse in Japan. Coastal species found in shallow weed and seagrass habitats to moderate depths in muddy substrates with algae reef outcrops. Variable, light to dark brown or yellowish with diffused pale blotches and banding for camouflage in weeds, or mainly blackish when in the open. Easily overlooked as it lives out in the open and is drably coloured to suit the surroundings, often covered with fine short filamentous algae (**B**), giving it a mossy look. Adults with a pale snout that blends in with the background. Has a smooth low crown. A small species with a height of about 12 cm, but tail long.

B

_H._ japonicus. Osezaki, Izu, Japan. Height 5 cm.

C D

H. japonicus. Osezaki, Izu, Japan. Height 7 cm. Hiroyuki UCHIYAMA.

E F G

H. japonicus? **E** Temperate Japan. Yutaka NIINO. **F** & **G** Ehime Pref, Japan. Gravid female, height 10 cm. Tomonori HIRATA.

Horned Seahorse
Hippocampus coronatus

Hippocampus coronatus Temminck & Schlegel, 1850. Nagasaki, Japan.

Japan endemic, Nagasaki to Tokyo Bay region. Weed habitats, originally found in floating weeds. Adults are identified by the tall backward-bend bony crown (**A**) and large flattened spines that point sideways from the dorsal fin base (**C**). Colour varies from light to dark red-brown, often with numerous thin whitish striations along the trunk and the head. Pouch of male often speckled with fine white and black spots. Height to 7 cm.

Remarks: species name is commonly used for *Hippocampus sindonis*, a species that is often seen by divers on reefs. *H. coronatus* is rarely seen by divers unless searched for.

H. coronatus. Male. Izu, Japan. Yasumasa KOBAYASHI.

B

C

H. coronatus. Oshima, Izu, Japan. Hisashi OHNUMA   H. coronatus. Female. (companion of male above). Izu, Japan. Yasumasa KOBAYASHI.

D

E

H. coronatus. Females in seagrass habitat. Tateyama Tiba Preference. Tokyo Bay, Japan. Usio Ide.

H. sindonis. Oshima, Japan. Hisashi OHNUMA.

Painted Seahorse *Hippocampus sindonis*

Hippocampus sindonis Jordan & Snyder, 1901. Totomi Bay, Hamamatsu, Japan.

Only known from sub-tropical, south-eastern mainland Japan. Lives in a range of habitats from shallow, high energy zone algae reef, to soft bottom habitats. Often spectacularly coloured with brilliant red or yellow. Females usually with fleshy extensions from spines above the eye and coronet. Males often with smooth coronet. Height to about 6 cm.

Remarks: *Hippocampus sindonis* & *H. coronatus* have 10 trunk rings, versus 11 in other Japanese species. *H. sindonis* is extremely variable in colour and the arrangement of fleshy tentacles on the head or trunk. It has a moderately high coronet that often has a large tentacle which has led to the confusion with *H. coronatus*. Bleeker described *H. mohnikei* from Kaminoseki Isl, Japan and this name has been applied to several species, including *H. sindonis*, but *H. mohnikei* is most similar to *H. comes*.

H. sindonis. Osezaki, Japan. Pair, female red and male (birthing) yellow. Hiroyuki UCHIYAMA.

H. sindonis. Osezaki, Japan. Individuals of pair above. Note diagnostic black chest spots. Hiroyuki UCHIYAMA.

H. sindonis. IOP, Izu Peninsula, Japan. Atsushi ONO.

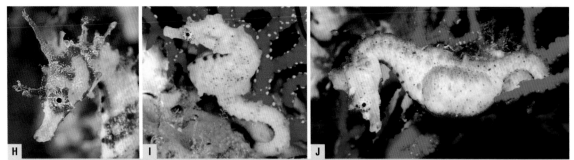

H. sindonis. Showing series of black spots ventrally on trunk. **H** Osezaki, Japan. Hiroyuki UCHIYAMA. **I J** Oshima, Japan. Hisashi OHNUMA.

Japanese Pygmy Seahorse *Hippocampus* sp 1

Presently only known from Hachijoh-jima Island, Izu Islands just south of Tokyo and Ogasawara Islands. Lives on mixed soft coral and algae reef and in shallow depths. There are a number of dwarf species that have only recently been discovered since divers are taking more notice of tiny creatures in the sea. Height 25 mm.

H. sp 1. Female. Hachijoh-jima, Japan. Shoichi KATO.

H. sp 1. Sub-adult. Ogasawara Islands, Japan. Osamu MORISHITA.

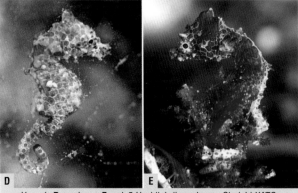

H. sp 1. **D** as above. **E** male? Hachijoh-jima, Japan. Shoichi KATO.

H. sp 1. Female. Hachijoh-jima, Japan. Shoichi KATO.

Hippocampus lichtensteinii Kaup, 1856. Probably Red Sea.

A Red Sea endemic and specimen in photographs were thought to be this species, but needs to be confirmed with a specimen check. Appears to be related to *H. bargibanti*. Skin tissue thick, hiding the ring-detail, and spine-like protrusion soft, equivalent to the wart-like ones of *H. bargibanti*. Lives in depths over 20 m, and height to 25 mm.

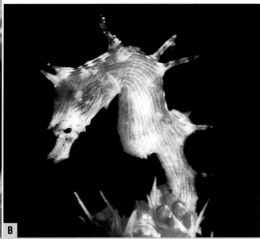

H. lichtensteinii(?) Sinai, Red Sea. Hans-Michael HACKENBERG.

Gorgonian Seahorse *Hippocampus* sp 2

At present only known from the included photographs taken in Milne Bay, Papua New Guinea, but maybe widespread and is very similar to the dwarf species from Japan (opposite page). Occurs on gorgonians from about 10 m down. It was thought to be the Pygmy Seahorse *H. bargibanti* but has different spine arrangements and male (**A**) has deeper trunk. *H. bargibanti* occurs in the same area on *Muracella* spp. and usually in much deeper water. It now seems that there maybe a number of species living on various gorgonians. Height 25 mm.

Should not be collected for aquarium purpose as it is a highly specialised species that probably can't live without its host.

H. sp 2. **A** male, **B** female. Milne Bay, Papua New Guinea, Height 25 mm. Bob HALSTEAD. **C** female. Roger Steene.

Borneo Pygmy-seahorse *Hippocampus* sp 3
Undetermined species

Appears to be a new species that was photographed by John Sear in Borneo. He found them in depths between 7 and 15 m. They were hiding in dark crevices on soft corals.

Remarks: similar *Hippocampus colemani* and the Japanese Pygmy Seahorse, but has a knob-like coronet and a flanged protrusion dorsally on the upper trunk.

H. sp 3. Borneo. John SEAR.

Coleman's Pygmy-seahorse *Hippocampus colemani*
Hippocampus colemani Kuiter, 2003. Lord Howe Island

This species is only known from Lord Howe Island and Milne Bay, Papua New Guinea. It was discovered by Neville Coleman (whom it will named after) at only 5 m depth in Lord Howe Island's main lagoon. Habitat comprises course sand with sparse *Zostera* and *Halophila* plants that have fine filamentous algae on their leaves. The same algaes are growing on the skin of the seahorse, giving them an exceptionally good camouflage and making them, together with their diminutive size, almost impossible to find. An unsualy feature is having only a single gill-opening that is situated just behind the top of the head that is surrounded by raised skin. This character appears to be shared by several pygmy-seahorse species. The largest of all the specimens observed was collected and measured 22.1 mm in height.

H. colemani. Lord Howe Island. Height about 20 mm. Neville COLEMAN.

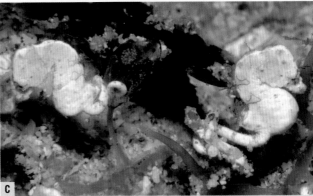

H. colemani. **B** Lord Howe Island. Neville COLEMAN. **C** Milne Bay, Papua New Guinea. Height about 12 mm. Roger STEENE.

Pygmy Seahorse
Hippocampus bargibanti

Hippocampus bargibanti Whitley, 1970.
New Caledonia.

West Pacific, Coral Sea, Australia, to southern Japan and ranging into Indonesia to Bali. It maybe expected in many other places. Highly specialised species that has adapted to life on various soft corals on which is feeds. Part of the coral appears to live symbiotically in the skin and reacts into warty-like growth around the short spines on the head and body. This gives a perfect camouflage and matches colour and shape of its host in detail, and may seem like different species. Post-pelagic young settle on various hosts, but to breed they appear to prefer the red-polyp *Muricella* spp that usually grows in depths over 20 m. A tiny species, height to 2 cm.

Little is known about behaviour of this species. I observed a small colony on a Gorgonian fan in Lembeh Strait, comprising at least 3 pairs and several more individuals. The 3 pairs were all very active one late October morning, males following the females and they interacted as shown in the photographs to the right. The male often pecked at the female, possibly making clicking noises like some other seahorse species do, but not loud enough for the human-divers ear. The males were all skinny and showed some evidence of a small pouch and by their behaviour seemed to be getting ready to take-on parenthood. The following year, in the same place, birth was observed by Denise Tacket and Sara Lourie. It was initiated accidently when Sara was trying to measure a pregnant male. It produced 34 tiny young that swim away to open water to start their initially pelagic life. They could travel far and is the reason for their large geographical range.

As this species is highly specialised, specimens should not be collected for aquarium purposes.

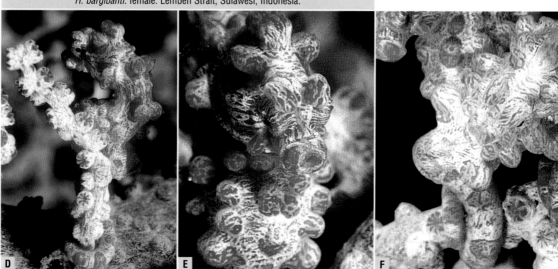

H. bargibanti. **A** male, **B** female. Lembeh Strait, Sulawesi, Indonesia.

H. bargibanti. female. Lembeh Strait, Sulawesi, Indonesia.

H. bargibanti. **D** Fish or Coral? **E** close-up of tubular mouth. **F** male, showing pouch-opening to female. Lembeh Strait, Sulawesi, Indonesia.

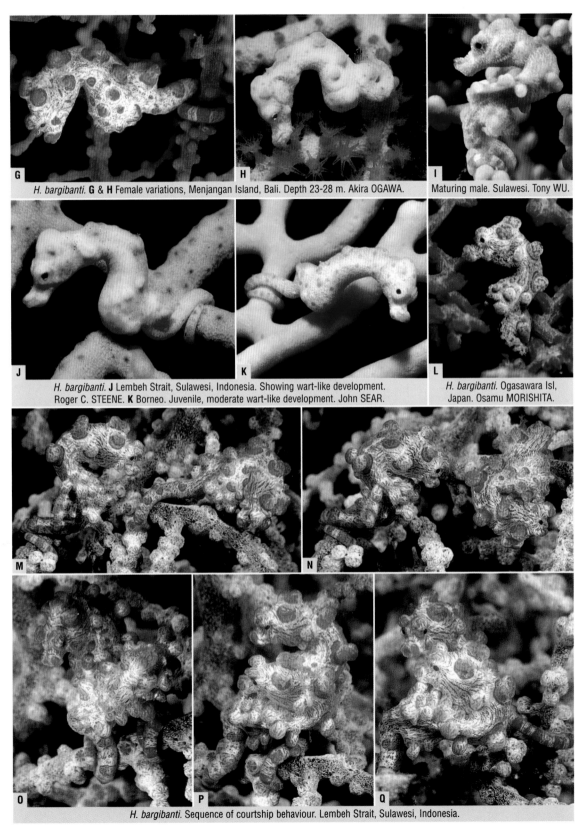

H. bargibanti. **G** & **H** Female variations, Menjangan Island, Bali. Depth 23-28 m. Akira OGAWA.

Maturing male. Sulawesi. Tony WU.

H. bargibanti. **J** Lembeh Strait, Sulawesi, Indonesia. Showing wart-like development. Roger C. STEENE. **K** Borneo. Juvenile, moderate wart-like development. John SEAR.

H. bargibanti. Ogasawara Isl, Japan. Osamu MORISHITA.

H. bargibanti. Sequence of courtship behaviour. Lembeh Strait, Sulawesi, Indonesia.

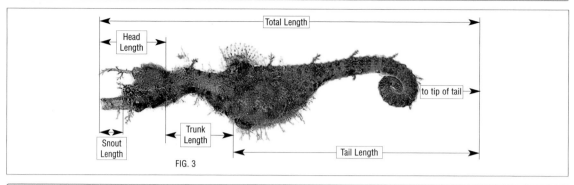

Total Length

Head
Length

to tip of tail

Trunk
Length

Snout
Length

Tail Length

FIG. 3

GENERAL

A group of small seahorse-like pipefishes, called pygmy pipehorses. They appear to be a link between the seahorses and pipefishes. The males have a moderately large tail pouch and causes them to posture more vertically than the female and makes them bend their head more to feed on the substrate. It is easy to see how seahorses have evolved that way. Their features are basically the same as the seahorses (FIG. 1 on page 10). Because the head is more in line with the body, the size is expressed in total length, from the tip of the snout to the end of the tail that may have to be unrolled (as above in FIG 3). The trunk and tail meet at the anus or pouch opening point. In the female this is more obvious compared to the male, especially if the male has a full pouch. The skin of these fishes attracts dermal-algae growth, which makes their camouflage perfect. They are extremely difficult to find during the day, but at night are sleeping near the top of the weeds.

GENUS *ACENTRONURA* Kaup, 1853

Feminine. Type species: *Hippocampus gracillissimus* Temminck & Schlegel, 1850. This genus comprises at least 3 species. The tropical West Pacific may have several more that are presently going under a single name. Females are much more slender than males and have a more pipefish like appearance, whilst the males of some species are more like seahorses with a deeper chest and a fully enclosing tail pouch. Small and highly camouflaged fishes that are locally common and found in small groups on soft-bottom with sparse algae.

A

Japanese Pygmy Pipehorse
Acentronura gracillissima
Hippocampus gracillissimus
Temminck & Schlegel, 1850. Japan.

Sub-tropical Japanese waters and records from elsewhere need to be verified. Occurs on rocky reefs in algae at various depths to at least 40 m. Often on open substrate with small rocky outcrops where they feed on mysids that swim over the sand. Variable in colour from almost black to purple or red. Females look very much like pipefishes. Length to 65 mm.

A. gracillissima. Female. Oshima, Japan. Length about 65 mm.

B

A. gracillissima. Osezaki, Japan. Length 45 mm. Hiroyuki UCHIYAMA.

C

A. gracillissima. Male. Osezaki, Japan. Hiroyuki UCHIYAMA.

Short-pouch Pygmy Pipehorse
Acentronura breviperula

Acentronura breviperula Fraser-Brunner & Whitley, 1949. Queensland, Australia.

West Pacific, but may comprise several species. Previously only a single species, *A. tentaculata*, was recognised, but this is a Red Sea endemic. Typically found on small and sparse seagrasses that grow adjacent to reefs. Shallow to moderate depths, and known range from 1–20 m. Usually found in pairs. Females are slender and pipefish like, but larger males are more seahorse like. Small species, usually to about 50 mm.

A

A. breviperula. Courting pair, Sulawesi, Indonesia. Length 50 mm.

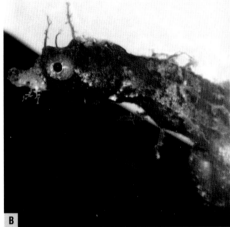

B

C

A. breviperula. Brooding male. Moreton Bay, Queensland. Depth 10 m. Length 50 mm.

D

E

A. breviperula. Brooding males. **D** Philippines. Length 40 mm. Roger C. STEENE. **E** Flores, Indonesia. Length 45 mm.

F

A. breviperula. Female. Flores, Indonesia. Length 40 mm.

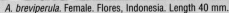

G

A. breviperula. Gravid female. Sulawesi, Indonesia. Length 45 mm.

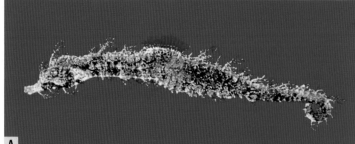

A

Acentronura tentaculata Günther, 1870. Gulf of Suez.

Red Sea endemic that may range to Arabian Seas and possibly West Indian Ocean (*Syngnathoides algensis* Fourmanoir, 1954, Commore Is, maybe this species). Variably coloured from green to brown and to near black. Usually with numerous hairy filaments when in algal beds. Mainly known from algae and sparse seagrass habitats in sheltered bays. Length to about 55 mm.

A. tentaculata. Disturbed from its habitat. Aqaba, Red Sea. Thomas PAULUS.

B

A. tentaculata. Brooding male. Aqaba, Red Sea. Thomas PAULUS.

C

A. tentaculata. Aqaba, Red Sea. Thomas PAULUS.

D

A. tentaculata. Female. Aqaba, Red Sea. Thomas PAULUS.

E

A. tentaculata. Aqaba, Red Sea. Thomas PAULUS.

Mozambique Pygmy Pipehorse
Acentronura mossambica

Acentronura mossambica Smith, 1963. Mozambique.
?*Syngnathoides algensis* Fourmanoir, 1954.
Commore Is. (Possible older name).

East African coast. Maybe more widespread in the West Indian Ocean. Mainly known subtidal zones in estuaries where among weeds. Previously included with *A. tentaculata*, but differs in having a longer brood pouch in males, growing larger, and in colouration. Length to about 63 mm.

A. mossambica. Brooding male 60 mm. Inhaca, Mozambique. Phil HEEMSTRA.

GENUS *AMPHELIKTURUS* Parr, 1930

Masculine. Type species: *Amphelikturus brachyrhynchus* Parr, 1930 (= *A. dendriticus*). Only 2 species known. Closely related to *Acentronura* and previously treated as subgenus. Studies on related genera suggest close relationship to the seahorses *Hippocampus* but amongst the pigmy pipe-horses, this genus is the most pipefish-like. Difficult to find and rarely observed fishes, associating with algaes.

West Atlantic Pygmy Pipehorse
Amphelikturus dendriticus
Siphostoma dendriticum Barbour, 1905. Bermuda.

Reported from south Florida to Gulf of Mexico and is probably widespread throughout the region. A shallow water species that is found on algae reefs to about 15 m depth, but because of its small size and extremely good camouflage it is rarely seen. Sometimes collected with floating *Sargassum* weeds. Length to about 50 mm.

The second undescribed species in the genus is only known from pelagic specimens from the east Atlantic.

A. dendriticus. Female. Little Cayman. Carl L. CAMPBELL.

GENUS *IDIOTROPISCIS* Whitley, 1947

Masculine. Type species: *Acentronura australis* Waite & Hale, 1921. Australian subtropical genus with 3, small, seahorse-like species. Difficult to find and rarely observed fishes, some only known from type-material. Differs from *Hippocampus* in head at less angle to body and dorsal fin with many more rays. Previously included with *Acentronura* as a subgenus but additional material shows closer affinities to *Hippocampus*.

Southern Pygmy Pipehorse *Idiotropiscis australe*
Acentronura australe Waite & Hale, 1921. South Australia.

Only known from a few specimens from Cape Jervis, St. Vincent Gulf, and Carnac Island, WA. Morphology similar to *I. lumnitzeri*. Colour unknown, probably matching algae habitat.

A

B

I. australe. **A** female. Perth, WA. **B** female above, male below. Cape Jervis, SA. After Waite & Hale.

Helen's Pygmy Pipehorse *Idiotropiscis larsonae*
Acentronura (Idiotropiscis) larsonae Dawson, 1984.
Alpha Island, Western Australia.

Only known from a single pair found clinging to attached *Sargassum* weed in 3–9 m at Monte Bello Isl. Length to 40 mm.

I. larsonae. Male, holotype. Monte Bello Is., WA. Length 33.5 mm. After DAWSON.

Sydney's Pygmy Pipehorse
Idiotropiscis lumnitzeri

Idiotropiscis lumnitzeri Kuiter, 2003.
Sydney, Australia

Only known from central New South Wales, Sydney to southern New South Wales. A spectacular little fish that is highly camouflaged amongst the short red algae on rocks. It has a special outer skin that encourages growth of the algae it lives amongst, often covering the rings and plates of the body. The leafy bits grow in particular places as part of the skin and somehow the shapes are transmitted as a code through the food-chain. Small crustaceans feed on the algae and pipehorses eat the small crustaceans, and the colour and shape of the algae comes out in the skin. This species is much more like a seahorse than a pipefish in all aspects, looks, morphology and behaviour. Males have a large pouch an estimated 60 eggs per brood. Occurs on semi-exposed open coast reefs in 6–30 m depth, and observed solitary, in pairs or small groups at night. Sydney diver Matthew Brooke observed and photographed this species over an 8-year period. He found that the various individuals live on the same small reef-section for long periods, some were monitored for more than 8 months. Length to about 55 mm.

A

I. lumnitzeri. Cronulla, Sydney, Australia. Pair, length about 45 mm. Matthew BROOKE.

B

I. lumnitzeri. Male, length 55 mm. Sydney, Australia. Stuart HUMPHREY.

C D

I. lumnitzeri. Females, Sydney, Australia. Matthew BROOKE.

E F

I. lumnitzeri. Brooding males, Sydney, Australia. Matthew BROOKE.

I. lumnitzeri. Cronulla, Sydney, Australia. Male with two females, length about 40 mm. Matthew BROOKE.

I. lumnitzeri. Cronulla, Sydney, Australia. Pair, length about 45 mm. Matthew BROOKE.

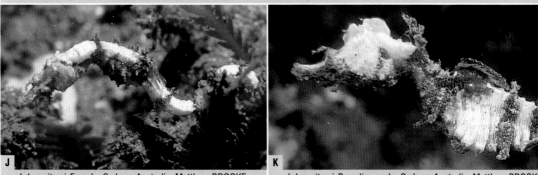

I. lumnitzeri. Female, Sydney, Australia. Matthew BROOKE.

I. lumnitzeri. Brooding male, Sydney, Australia. Matthew BROOKE.

I. lumnitzeri. Female, 35 mm. Sydney, Australia. Matthew BROOKE.

Comprises 5 genera, *Solegnathus* with 9 species, others with a single species each. A diverse group, of which 3 of the genera are Australian endemics, the seadragons. *Solegnathus* comprises the pipehorses, the genus that has representatives distributed throughout the West Pacific and only the closely related genus, *Syngnathoides* is widespread through the Indo-West Pacific. All species lack a caudal fin and the tail is prehensile in the pipehorses. Males carries the brood exposed under the tail, except for *Syngnathoides* where under the trunk. Eggs are large and usually number about 300. They don't have pelagic stages, except for *Syngnathoides,* which floats in sargassum rafts up to adult size.

Diet comprises various crustaceans and larval fishes.

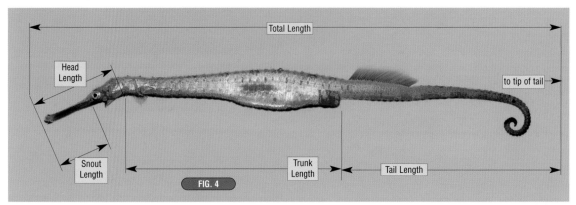

FIG. 4

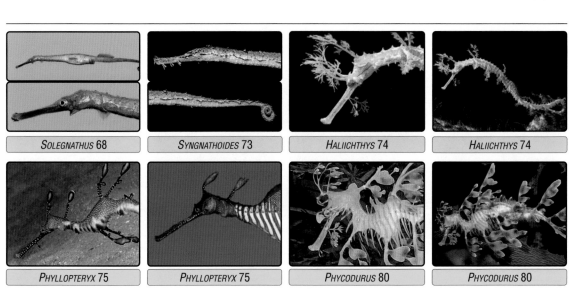

| *SOLEGNATHUS* 68 | *SYNGNATHOIDES* 73 | *HALIICHTHYS* 74 | *HALIICHTHYS* 74 |
| *PHYLLOPTERYX* 75 | *PHYLLOPTERYX* 75 | *PHYCODURUS* 80 | *PHYCODURUS* 80 |

GENUS *SOLEGNATHUS* SWAINSON, 1839

Masculine. Type species: *Syngnathus hardwickii* Gray, 1830. At least 9 species occur in the West Pacific, but those reported as the same species from Japan and Australia, or those from tropical and sub-tropical regions, should be investigated. The presently recognised species are known from Australia, one reported from New Zealand, one from Japan, and two from Indonesia. Because of their breeding methods and lack of pelagic stages, these fishes are localised, but due to their deep water preference morphological changes are minimal over evolutionary times. They are much like straightened seahorses with spiny rings. The dorsal fin is large and placed over the tail, just behind anal origin. In Australia and New Zealand they are known as spine pipehorses, but in Asian countries they are usually called spiny seadragons. This has led to confusion in the trade of dried specimens for the Chinese Traditional Medicinal market, and erroneous claims that the protected Australian seadragons are used for this. However, moderate quantities of *Solegnathus* are used, most locally caught, and few are exported from Australia where they are a bycatch in trawls in depths of about 100–150 m. Most species live on open soft -bottom substrates adjacent to reefs where soft-corals and seawhips are scattered in current prone areas. Because of their habitat near reefs, they are generally avoided by trawlers in Australia.

Hardwick's Pipehorse
Solegnathus hardwickii
Sygnathus hardwickii Gray, 1830. China.

China Seas to southern Japan, and two similar species in northern Australia, one each on the western and eastern coasts. May comprise several more species. True *S. hardwickii* from northern hemisphere is white with black markings, forming lines along back ridges from behind head to almost tip of tail. Western Australian form has dark marking over the upper ridges to dorsal fin and tail often barred. The marking on the back appear to serve as a mimic for seawhips and with its head down, keeps its back towards potential threat (**A**). Australian form(s) undescribed. Reported from trawls in less than 100 m, but enters relatively shallow depths of about 40 m. Length to about 50 cm.

S. hardwickii. Okinoshima, Kochi Preference, Japan.
A 'headstand' pipehorse, a perfect mimic of a seawhip. Takao OKADA. **B**

C

S. hardwickii. Trawled specimen, Kashiwajima, Kochi Preference, Japan. Length 45 cm. **B** dorsal view.

D

S. hardwickii. Okinoshima, Kochi Preference, Japan. Depth about 40 m. Takao OKADA.

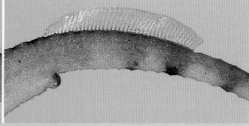

Western Pipehorse *Solegnathus* sp 2

Undescribed species that was previously included with *S. hardwickii*. Short dark line through the eye, lower trunk yellow and cheek yellow with white streak. Tail of female broadly barred with diffused dusky bands and ventral pale patches. Back ridges with dark spots, variable from few to almost forming lines. In depths of about 100 m or more. Length to about 50 cm.

A

S. sp 2. Female. NE of Cape Lambert, WA. From 95 m. G. YEARSLEY, CSIRO FISHERIES (sections from curved specimen).

B

S. sp 2. Male showing depressions on the tail from a brood. Off Port Hedland, WA.. From 132 m. Length 45 cm. T. CARTER, CSIRO FISHERIES.

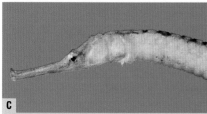

C

S. sp 2. Male. North West Shelf, WA. From 100 m, CSIRO FISHERIES (sections from curved specimen).

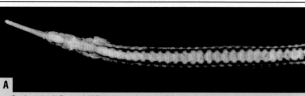

A

S. dunckeri. Central NSW. Dorsal view of male. L. 38 cm. After DAWSON.

Dunker's Pipehorse *Solegnathus dunckeri*

Solegnathus dunckeri Whitley, 1927. Lord Howe Island.

Only known from central New South Wales to southern Queensland. Trawled between 75 and 140 m depth. Has black stripe along back ridges and dark tail. Was put in separate sub-genus on the basis of trunk to tail ridges being slightly different in their way of connecting. Length to about 50 cm.

B

S. dunckeri. Northern NSW. From 40-50 m depth. Female. L. 45 cm. Ken GRAHAM.

Robust Spiny Pipehorse
Solegnathus robustus

Solegnathus robustus McCulloch, 1911.
Off Flinders Island, South Australia.

Only known from the south Australian region where trawled in less than 100 m depth. Replaces *S. spinosissimus* west of Bass Strait. It differs from that species in its shorter and thicker snout, and has a less oval shape of the tail cross section. A plain pale species that reaches about 40 cm in length.

S. robustus. Flinders Island, South Australia. Museum specimen, 40 cm long.

Queensland's Spiny Pipehorse
Solegnathus sp 1

Undescribed species that was previously included with *S. hardwickii*. Known from off Queensland and northern New South Wales. Yellow line from about centre of snout through the eye to end of head, cheek white below line and additional short yellow line below from eye. Series of black spots along back ridges. Snout very long. Usually trawled near 100 m depth. Length to about 50 cm.

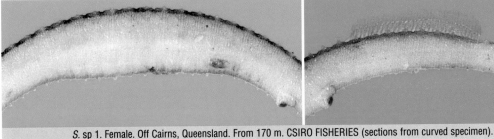

S. sp 1. Female. Off Cairns, Queensland. From 170 m. CSIRO FISHERIES (sections from curved specimen).

Indonesian Pipehorse
Solegnathus lettiensis

Solegnathus lettiensis Bleeker, 1860.
Letti Island, Banda Sea, Indonesia.

Indonesia, Moluccen, Flores and Java Seas. Has large saddle-like markings over the back. Little known species, probably reaching 45 cm.

S. lettiensis. Banda, Indonesia. After BLEEKER.

Günther's Pipehorse
Solegnathus guentheri

Solegnathus guentheri Duncker, 1915.
Houtman Abrolhos, Western Australia.

Sub-tropical Western Australia from its southern tip to the Houtman Abrolhos. Known from few specimens, trawled in about 150–180 m depth. Usually has several large black blotches in the sides. Length to 52 cm.

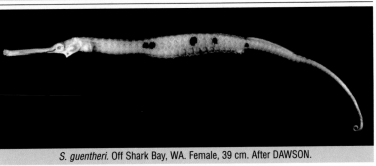

S. guentheri. Off Shark Bay, WA. Female, 39 cm. After DAWSON.

71

Australian Spiny Pipehorse
Solegnathus spinosissimus
Solegnathus spinosissimus Günther, 1870. Tasmania.

South-eastern Australia, from Bass Strait region and Tasmania, just ranging into southern Queensland. Occurs on soft-bottom in depth over 30 m to 250 m, usually rubble substrate, and near rich invertebrate platform reefs. Only seen in shallower depths in the southern part of its range in estuaries that are shaded or the water darkened by tanning. By-catch in trawls and now exported as dried specimens for Chinese Traditional Medicine. Before that specimens were washed over the side and died as they were unable to go back to the bottom with their swimbladder inflated. These were often washed up on the beach after strong on-shore winds. A large species that can reach 49 cm in length.

S. spinosissimus. Derwent Estuary, Tasmania. Male with partly hatched brood.

S. spinosissimus. Specimen from off Lake Entrance, eastern Victoria. Length 32 cm.

New Zealand Spiny Pipehorse
Solegnathus naso
Solegnathus robustus naso Whitley, 1941. Auckland fish market, New Zealand.

Widespread New Zealand, previously regarded the same as *S. spinosissimus*, but differs in colour and has a more slender snout. Enters shallow waters in southern New Zealand. Young are benthic, and, like other congeners, have no pelagic stage. Length to 50 cm.

S. naso. **A** Fiordland, New Zealand. Depth 15 m. K.R.GRANGE.**B** Close-up of brood. South-eastern NZ, from 15 m depth. Malcolm FRANCIS

S. naso. Brooding male. Fiordland, Southern New Zealand. Depth 25 m. Mary MALLOY.

Masculine. Type species: *Syngnathoides blochii* Bleeker, 1851 (= *Sygnathus biaculeatus* Bloch, 1785). Represented by a single widespread shallow water species. Floats with sargassum rafts after wet-seasons that enabled it to disperse over great distances.

Double-ended Pipehorse *Syngnathoides biaculeatus*

Sygnathus biaculeatus Bloch, 1785. East Indies.

Widespread tropical Indo-West Pacific from east Africa and Red Sea to Japan and central Pacific. Usually found in sheltered lagoons amongst broad-leafed seagrasses. Large numbers of adults were observed in Maumere Bay after the wet-season in sargassum rafts. When they became alarmed from the approaching camera, some jumped on top of the weed to escape potential danger from below. This technique was also observed with the sargassum anglerfish *Histrio histrio*. They would stay out of the water for a considerable length of time, before jumping back in. Males carry the exposed brood underneath the trunk (**C**). Variable green to brown or grey, depending on habitat. Females are ornamental below the trunk that is used when she courts the male. Length to about 28 cm.

A
S. *biaculeatus*. Maumere Bay, Flores, Indonesia. At surface amongst *Sargassum* raft.

B
S. biaculeatus. Bali, Indonesia. Female, 25 cm.
Takamasa TONOZUKA.

C
S. biaculeatus. Sanur Lagoon, Bali, Indonesia. Male with brood. Length about 25 cm.

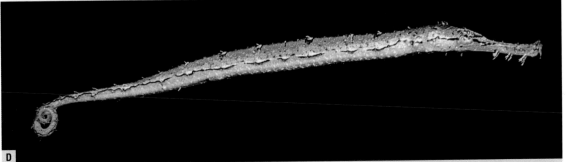

D
S. biaculeatus. Maumere Bay, Flores, Indonesia. Surface waters amongst *Sargassum* raft. Length about 25 cm.

Masculine. Type species: *Haliichthys taeniophora* Gray, 1859. Represented by a single tropical shallow water species. Related to the southern genus *Phycodurus* and appeared to have changed least since their ancestral form.

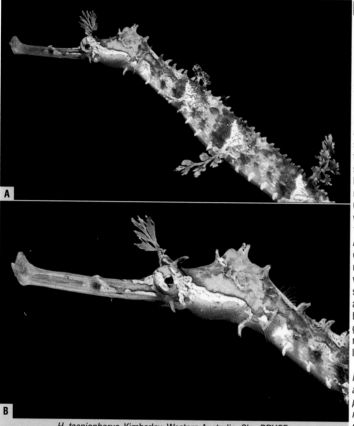

A

B

H. taeniophorus. Kimberley, Western Australia. Clay BRYCE.

Ribboned Seadragon *Haliichthys taeniophorus*

Haliichthys taeniophora Gray, 1859. Shark Bay, Western Australia.

Primarily known from the north-western Australian coast, ranging north from Shark Bay to Exmouth, but also known from the Darwin region and east to Torres Strait, including the New Guinea region. Reports by Pearl divers of a leafy seadragon in the Broom region had puzzled the author for a long time as to what species it could be. A photograph finally sent by Underwater World in Perth confirmed that there was indeed a third seadragon species. It showed this species which until then was mainly known from preserved trawled specimens, usually missing the leafy appendages. It was named by Gilbert Whitley as the Ribboned Seadragon, but it was later published as ribboned pipehorse (Munro, 1958) and as ribboned pipefish (Dawson, 1985 & Allen & Swainston, 1988), latter illustrated a painting of a specimen without any leafy appendages that nobody could recognise. It lives mainly in shallow water in weedy zones bordering open substrates such as tidal channels to depths of about 16 m but also deeper on soft-bottom substrates. Colour is highly variable, shallow water specimens mainly greenish yellow, but trawled specimens are brown to reddish with large blotched patterns. Reaches a length of at least 30 cm.

No doubt an interesting species that has a lot of aquarium potential and should be bred for this purpose. It is more tropical than its southern relatives and therefore more suitable for many public aquariums

C

H. taeniophorus. Photograph courtesy Underwater World, Public Aquarium, Perth, Western Australia.

Masculine. Type species: *Syngnathus foliatus* Shaw, 1804 (=*Syngnathus taeniolatus* Lacepède, 1804). Represented by a single sub-temperate shallow water species, some geographical variation. Australian endemic.

Weedy Seadragon *Phyllopteryx taeniolatus*

Syngnathus taeniolatus Lacepède, 1804. New Holland (=Australia).

Widespread in southern Australian waters from the Sydney region on the east coast to the Perth region on the West coast to the southern tip of Tasmania. Several species were proposed for different geographical locations, but the differences are based on variations that are normal in wide-ranging species. Southern Tasmanian population grows much larger and have deeper bodies compared to those from the northern end of their geographical range which is to be expected. Colour and leafy appendages are influenced by environment and food that can vary considerably with depth and geography. Body deepens with age, especially in females. Specimens from deep water are brighter coloured and less 'leafy'. No doubt that some forms have localised features or colours as they are non-pelagic. They are solitary species that behave differently in the various geographical zones. They usually have only one brood per season, and in principle their mating starts in October and November. In some areas this can vary and with favourable conditions may have a second brood. Large males have about 250-300 eggs per brood that is openly visibly below the tail (**A**). Each egg is embedded in the skin that was soft when they were laid and reacted by raising around each one and hardened, providing a cup to hold them secure. Young hatch after about 2 months and settle on the substrate. Their snout is short and they lack the elaborate leafy bits, but this grows quickly and after two days their snout is produced whilst their yolk that supported them has run out, but they are ready to hunt. In Victoria they are often abundant in very shallow weedy areas of Port Phillip Bay and Western Port in the areas that are adjacent to the tidal channels, usually in *Ecklonia* dominated low reefs. In New South Wales they are more common in depths over 15 m on the sheltered reefs of the open coast but shallower in large open estuaries such as Jervis Bay. They are often locally common and spread out over reefs, usually where food is abundant. In some areas they follow food and may be in large numbers on occasion. Some move to deeper water during the winter months when food is scarce. Length to 45 cm.

A

P. taeniolatus. Male with brood. Bermagui, NSW.

B

P. taeniolatus. Portsea, Port Phillip Bay, Victoria. They are easily approached if not harassed, otherwise they become shy of divers.

This species is exported to public aquariums around the world. Most specimens are tank-raised in Victoria which have proven to be the most successful. Wild adults do not easily adapt to captivity and are best left alone as they require a lot of space and transport has a low rate of success. Tank-raised individuals are usually send at less than half-grown size and have a 100% transport survival rate. In addition they are used to the aquarium conditions and readily adapt to more restricted space or disturbances. They grow quickly and in one year are almost fully grown. Specimens can live well over 10 years in captivity and breeding is possible in high tanks which is needed for their egg-transfer as they usually rise towards the surface to do this. In the wild this maybe several meters and if they hit the surface in the aquarium they may panic and drop the eggs. This has happened on several occasions but some public aquariums in Japan have been successful in producing and growing young.

P. taeniolatus. Close-up of head, showing very long snout with small mouth at tip. Portsea, Vic.

P. taeniolatus. Female. Flinders, Westernport, Victoria, Australia.

P. taeniolatus. Brooding male. Flinders, Westernport, Victoria, Australia.

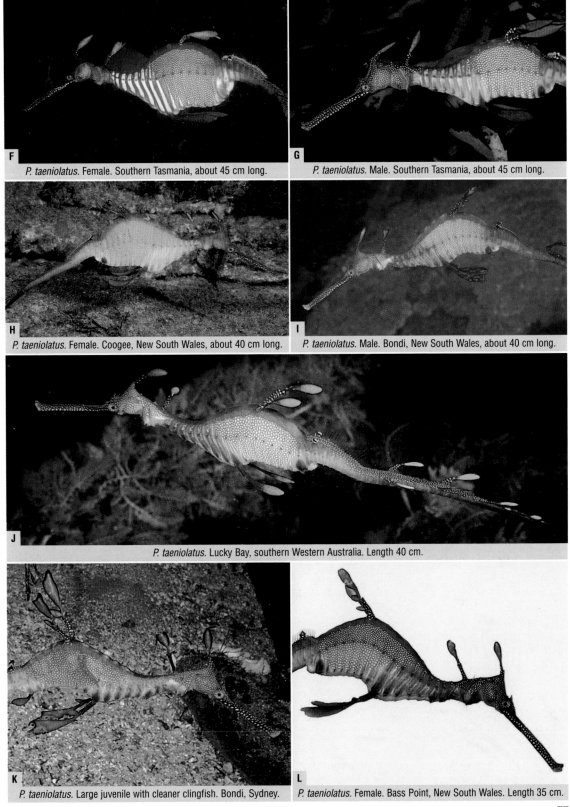

F   *P. taeniolatus.* Female. Southern Tasmania, about 45 cm long.

G   *P. taeniolatus.* Male. Southern Tasmania, about 45 cm long.

H   *P. taeniolatus.* Female. Coogee, New South Wales, about 40 cm long.

I   *P. taeniolatus.* Male. Bondi, New South Wales, about 40 cm long.

J   *P. taeniolatus.* Lucky Bay, southern Western Australia. Length 40 cm.

K   *P. taeniolatus.* Large juvenile with cleaner clingfish. Bondi, Sydney.

L   *P. taeniolatus.* Female. Bass Point, New South Wales. Length 35 cm.

M

P. taeniolatus. Victor Harbour, Southern Australia. 3 juveniles, blending in with the background. Although in large numbers in January, they are rarely noticed.

N

P. taeniolatus. Hatchling, just born. Victoria.

O

P. taeniolatus. Hatchling, one day old. Victoria.

P

P. taeniolatus. Flinders, Vic, Australia. Tiny juvenile in centre.

Q

P. taeniolatus. Hatchlings, few days old. Victoria.

78

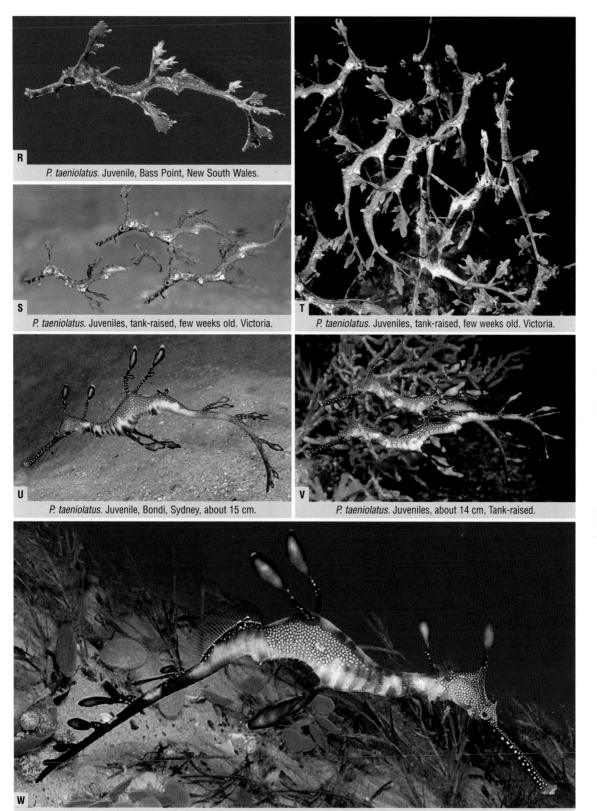

R P. taeniolatus. Juvenile, Bass Point, New South Wales.

S P. taeniolatus. Juveniles, tank-raised, few weeks old. Victoria.

T P. taeniolatus. Juveniles, tank-raised, few weeks old. Victoria.

U P. taeniolatus. Juvenile, Bondi, Sydney, about 15 cm.

V P. taeniolatus. Juveniles, about 14 cm, Tank-raised.

W P. taeniolatus. Large juvenile, Victor Harbour, South Australia. About 20 cm.

Masculine. Type species: *Phyllopteryx eques* Günther, 1865. Represented by a single sub-temperate shallow water species, which is an Australian endemic.

Leafy Seadragon *Phycodurus eques*

Phyllopteryx eques Günther, 1865. Port Lincoln, South Australia.

Common in South Australia (the State) and southern part of Western Australia, with sporadic occurrence in southern Victoria and the southern west-coast of Western Australia, rarely seen north of Perth. Reports from tropical waters are based on the Ribboned Seadragon *Haliichthys taeniophorus*. Primarily found along edges of *Ecklonia* dominated reefs bordering onto sand where they feed on mysids and shrimps, depending on their size. Large adults learn to target various reef crustaceans, including large shrimps and squat-lobsters. Adults may congregate in certain shallow bays in late winter to pair and eventually mate. At other times the majority live in the deeper parts where *Ecklonia* is struggling to survive because of lack of light. In off-shore waters where visibility is greatest, these fishes occur in loose groups in depths of about 25 m. Shallow coastal populations fluctuate during the year and are influenced by freshwater run-offs and rough conditions. Most specimens are observed by divers in shallow depths, as in South Australia, most deeper off-shore reefs are not popular to dive since Great White Sharks are common visitors there. In the Esperance area, WA, the islands have large populations of Leafy Seadragons that are mostly seen over 20 m depth. Deep water individuals are dark brown to burgundy red in colour, whilst shallow water ones are more yellow or greenish. One of the most popular dive-sites for these fishes is Victor Harbour, S.A., because of easy access, but populations fluctuate and some areas have sanded in over the years. Other areas are coming known with local dive-operators finding specimens in 'their' areas. Specimens are often harassed by curious divers who can't resist touching them, or by impatient photographers. This causes them to move somewhere else. To get good photographs, it is important to stay at a distance that doesn't concern the seadragon. Let it come to you, but never chase. If closing in on the subject, at the first sign of concern by it, back-off and slowly try again. This may take some time, but it will be worth it.

Whilst this species is abundant in some areas, they are easily missed by divers, even when looking out for them. I will never forget when I noticed a large *Isopod* on a 'weed', that swam away - a leafy seadragon. On several occasions I've taken experienced fish-watchers from NSW and Victoria to Victor Harbour to look at the leafies, and always had to point-out the first one before they could discover one of their own. On one occasion under Rapid Bay jetty, a photographer couldn't see a specimen feeding out in the open about a metre in front of him, even when pointing it out. Only after making the leafy swim did it click! I vividly remember seeing my first Leafy Seadragon over 25 years ago. It remains my all time favourite creature in the sea, and as I wrote once before, is:

'the ultimate in evolution'

Without stretching the body parts into line, it grows to about 35 cm.

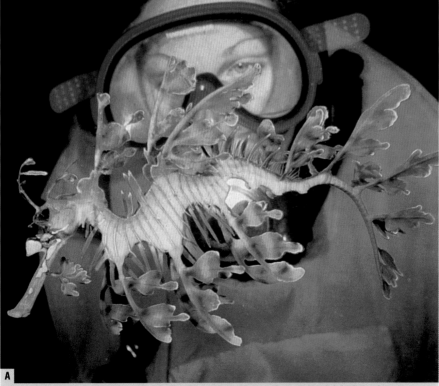

This species is one of the most rewarding for public aquariums. Wild adult usually suffer during transport, and are very sensitive to light changes, in addition are difficult to adapt to captivity. Aquarium raised specimens are usually exported and these can live in excess of ten years, depending on conditions in captivity. They are used to people, and captivity conditions, such as artificial lights.

When raised from eggs, they grow rapidly for the first few months to almost half adult-length. Growth rate slows but weight increases at this stage and at nearly one year of age they are small adults, ready to breed. In two years they reach full size, growing very little, if at all, during the next few years. They possibly live well over 10 years, and will breed if given space, especially height required for spawning (the higher, the better. 1.5 m recommended minimum).

A
P. eques. Esperance, Western Australia. Diver observing a perfect young adult leafy seadragon.

B

C

P. eques. Victor Harbour, SA. **B** Close-up, showing the large bony protrusion in front of the eyes probably serves as an aim for prey. **C** Deep-water colouration. Reports by fisherman of red dragons proved to be true when finding this specimen at 30 m.

D

P. eques. Rapid Bay, SA. Photographed with flash, the seadragon shows up as it lights-up the sides of the fish. With natural light it is lost in the background, and divers not aware of the fish in the area will not see it. Even when looking for this species, few are usually found and if not watched carefully it can quickly disappear amongst the vegetation. If not harrassed by divers, these fishes can be approached at close range, and if watched for a while it may swim towards the diver and feed right in front. If showing any concerns it is best to move back until it settles down. After a while it will realise that there is nothing to worry about and will continue its business that is mostly looking for food. They should never be chased, and especially not handled, as this will spoil it for the next diver.

Phycodurus eques

'ULTIMATE IN EVOLUTION'

E

P. eques. Victor Harbour, South Australia. Brooding male, 28 cm long.

PREGNANT MALES

Males brood an estimated 250–300 eggs, covering most of the under and sides of the tail, immediately behind the anus. They are partly embedded in the skin, about 4 mm in diameter and at best estimate about 7 mm long. Pregnant males are usually seen during the months of November and December, but this may vary seasonally. After mating many move to deeper water, depending on the area. Gravid females and males that are getting ready to spawn may congregate in certain areas in October. Prior to mating the tail of the male looks swollen and maybe bright yellow. It seems to be conditioned for the purpose, holding the sperm to meet the eggs. The female deposits her eggs in a similar way shown in the pipefish sequence (see P. 110), in a single sheet as they rise from the substrate. Each egg needs to be pushed onto the skin to become fertile. The sperm is not water-born and how it travels on or in the skin is not known, but ducting has been found in the pouch laminae of seahorses that in principal spawn the same way. Eggs that fall away, as sometimes happens in shallow aquariums, are never fertile, but would be if sperm were free-swimming. When eggs are deposited, the soft skin reacts and rises around each individual egg and hardens, holding it secure. The incubation time is about 8 weeks and hatching young are well advanced, settling and swimming close to substrate. Young are about 35 mm long and unroll from their tight position in the egg capsule by coming out tail first, taking about 3–6 hours before ready to swim. They hatch with a few individuals at the time and over a long period of 6–7 days. This way, the male can distribute its young over a wide area. The young may travel to shallower depths from their birth place. They are often seen in shallow sand and weed and rubble mix in sheltered places in about 5–7 m depth, sometimes in small loose aggregations.

P. eques. Victor Harbour, South Australia. Brooding male, 24 cm long.

P. eques. Victor Harbour, South Australia. **H** female left, male right, near spawning **I** Gravid female, ready to spawn. **J** Male, ready to receive brood.

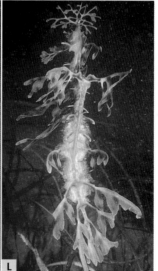

SPAWNING

Just prior to spawning the two sexes congregate in some areas and females have greatly swollen trunks (**I**), whilst males are much more slender (**J**), but these also show their readiness by their wrinkled area along the ventral and side of the tail. This maybe a reaction of the sperm making its way over the section that holds the brood. The skin appears to be soft and spongy. After the eggs are laid the skin hardens around the lower part of the egg and forms cups.

P. eques. Victor Harbour, South Australia. Brooding male. In the shallows, algae grows on the eggs in a few weeks.

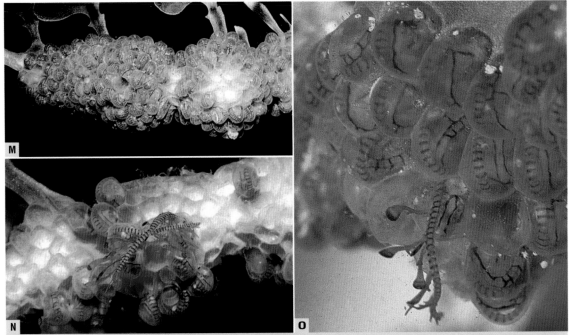

P. eques. Victor Harbour, South Australia. Hatching eggs. **M** young are fully coloured, darkening the colour of the eggs. **N** hanging out of eggs to recover from the tightly rolled-up position in the egg, and egg-pockets left in the skin where empty ones have dropped off. After all young are gone, the skin recovers. **O** young appear tail first and probably cut the hole with the spines along the end of the tail. The hole is shown in the empty egg to the right.

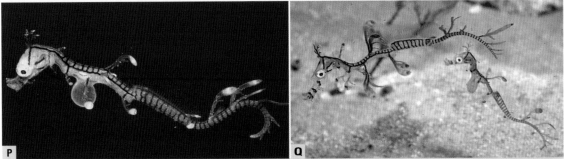

P. eques. **P** hatchling in aquarium, just minutes old, with yolksac that will support it for two days, in which time it snout is produced and appendages expand. About 35 mm long, but after a few more days it grows rapidly as shown in **Q** where it can be compared with a newly born young below its tail.

P. eques. Victor Harbour, South Australia. In January many juveniles occur in sheltered shallow reefs on small sand patches where they feed on small mysids. Often two or three are found together, those in the photographs only a few days old, in 7–8 m depth at the Bluff.

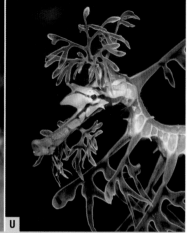

P. eques. Aquarium grown specimens at about 90 mm long after 4 weeks.

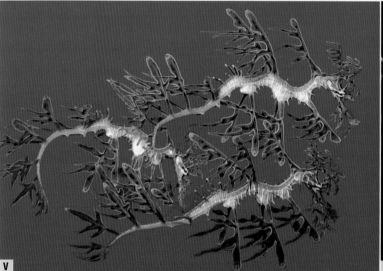

P. eques. Aquarium grown specimens at about 140 mm long after 14 weeks.

P. eques. On display in the Singapore Aquarium 'Underwater World'.

SEADRAGONS IN THE AQUARIUM.

At 2 years of age, aquarium grown specimens look spectacular (**X**), and have elaborate and undamaged appendages. The leafy seadragons are now on display in many aquarium around the world and this has brought numerous divers to South Australia to see them in the wild. Most specimens originated from aquarium grown specimens, and since several years only those hatched and grown in aquarium are exported. This way there is no threat to any wild populations and aquarium produced specimens can be shipped at smaller sizes and are used to captivity. They grow in two years to full size. The initial growth is fast, to about 200 mm in less than one year, but as in the wild, their growth-rate slows dramatically after that period.

Y

Z

P. eques. Aquarium grown specimens about 90 mm long and 4 weeks old.

Z1

Z2

Cleaner Shrimp *Periclimenus aesiops*. This species is closely related to the tropical shrimps that live in anemones, and the only known member of that group found in southern sub-temperate waters. It is often seen on ascidians and certain fishes may visit the site to get rid of irritating parasite that maybe a delicacy to the shrimp. Here it swam onto the leafy seadragon at Victor Harbour, SA. Seadragons also visit cleaner clingfish but these are generally in less suitable or accessible areas for the dragons.

P. eques. Mondrain Island, Recherche Archipelago, southern Western Australia. Juvenile, about 15 cm, and about 2 months old (photographed early March).

Specimens in this area varied from greenish yellow to golden brown, and some large adults in depths over 25 m were a deep red-brown. As their diet comprises various small crustaceans that primarily feed on algaes, the colour of dominant red in the deep and green in the shallows may carry through the food-chain, and the main factor in their general colouration. Aquarium specimens fed on red mysids from kelp areas become deeper coloured compared with those fed on pale mysids from *Zostera* dominated habitats.

Comprises 4 genera, *Doryrhamphus* the largest with more than 11 species, *Dunckerocampus* with 7 species, *Heraldia* with two species, and *Maroubra* with two species. Unlike most other pipefishes, the species are free-swimming and avoid the substrate, except *Maroubra* that lives loosely on clean rocks. They feature a large caudal fin that is usually brightly coloured and distinctively marked with stripes or bands, a pattern that is usually diagnostic for a species. Most species have been observed cleaning other fishes, apparently taking small copepods or other crustaceans. Adults commonly occur as pairs and live under cover of large coral pieces, in caves or overhangs of reef over sand. It is difficult to determine the sex for most species, except when the male incubates a brood, but adults usually occur in pairs of the opposite sex. The male incubates the brood under the trunk and the eggs are semi-exposed or covered by a skin membrane. Eggs are of moderate size and number from about 50–200. The young are well developed when hatching but have a pelagic stage for a short period, and dispersal distances are probably not great. Recent taxonomy, primarily based on preserved specimens, failed to differentiate between similar or closely related species that have a near identical morphology. This was especially the case in *Doryrhamphus* as many different species were going under the same name.

Diet comprises primarily tiny crustaceans that swim near the bottom, are picked from the substrate, or cleaned from other fishes and even from large crustaceans, and some target zooplankton. They are ideal fishes of small or primarily invertebrate aquariums, provided that live food can be provided on a regular basis. Most species will breed readily in captivity when given enough shelter and can live for about 5 to 10 years.

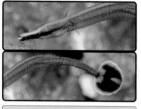

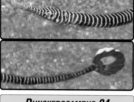

| DORYRHAMPHUS 88 | DUNCKEROCAMPUS 94 | MAROUBRA 100 | HERALDIA 101 |

Genus *Doryrhamphus* Kaup, 1856

Masculine. Type species: *Doryrhamphus excisus* Kaup, 1856. An Indo-West Pacific, comprising at least 11 tropical species, many of which were lumped together as single widespread species. Free-swimming benthic fishes with large tails that bare diagnostic colour patterns. They are found in various reef habitats in coastal to outer reefs, and usually stay close to small caves or narrow crevices in which they can retreat. Many species are known to be active cleaners, picking tiny parasitic crustaceans from other fishes, and adults work mostly in pairs. Males brood eggs semi-exposed under the trunk, sometimes a thin skin covering over the sides of the brood. About 80–150 eggs are incubated, depending on age or species. As tiny juveniles can be found on reefs, it seems that they have either a short or no pelagic stage. Most species have a restricted distribution and some have localised colour forms.

D. excisus. Oman. Depth 8 m. Length 65 mm. Phil WOODHEAD.

Indian Blue-stripe Pipefish
Doryrhamphus excisus

Doryrhamphus excisus Kaup, 1856.
Masawa, Red Sea.

Red Sea and western Indian Ocean, but may range to eastern parts of the Indian Ocean fauna. Similar to *D. melanopleura*, but caudal fin orange with black at base and usually a dark elongated spot from upper and lower centre to posterior margin with variable black central areas. Length to 55 mm.

Pacific Blue-stripe Pipefish *Doryrhamphus melanopleura*
Syngnathus melanopleura Bleeker, 1858. Kokos Isl, Indonesia.

West Pacific, Indonesia to southern Japan and Micronesia. Previously included with *D. excisus.* Blue stripe along upper sides is wide and lacks dark borders. Caudal fin variable and usually with elongated spot, centrally near its posterior margin. Males carry about 85 large eggs in a simple semi-open pouch under the trunk (**C**). Secretive under large coral pieces on shallow reef flats and lagoons, or in low caves where it usually swims upside down against the ceilings. Adults usually in pairs and pick parasites off other fishes. Length to 60 mm.

A

D. melanopleura. Philippines. Length 60 mm. Photo: R. C. STEENE.

B

D. melanopleura. Maumere, Flores, Indonesia. L. 50 mm.

C

D. melanopleura. Sanur Lagoon, Bali, Indonesia. Male, length 60 mm.

Barrier Reef Pipefish *Doryrhamphus* sp 1
Doryrhamphus sp.
Undescribed species from Queensland and Coral Sea.

Tropical eastern Australia and Coral Sea, ranging north to at least Papua New Guinea. Caudal fin orange with dark central area that is usually shaped as a 'C'. This species was confused with *D. melanopleura,* which has more black in the caudal fin. Occurs mainly in sheltered lagoons in inner reefs, usually in pairs under large dead coral slabs on rubble zones. Length to 55 mm.

A

Doryrhamphus sp. Lizard Island, Queensland. L. 50 mm.

B

Doryrhamphus sp. GBR, off Cairns, Queensland. L. 50 mm. Phil WOODHEAD.

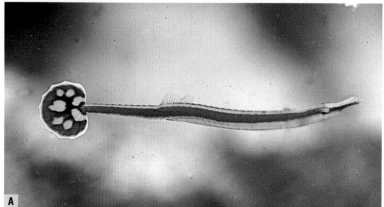

Flower-tail Pipefish
Doryrhamphus sp 2.

Probably restricted to the southern range of the Ryukyu Islands of Japan and Taiwan. Reported as *D. excisus*, now regarded as a Red Sea and West Indian Ocean endemic. Closely related to *D. japonicus* and *D. melanopleura*. Has broad blue band comparable to *D. melanopleura*, and distinctive caudal fin pattern with many yellow spots that are well defined. Secretive in sheltered reefs. Length to 60 mm.

D. sp 2. Iriomote Isl, Japan. L. 40 mm. Toshikazu KOZAWA.

D. sp 2. Iriomote Isl, Japan. L. 60 mm. Korechika YANO.

Honshu Pipefish
Doryrhamphus japonicus

Doryrhamphus melanopleura japonica Araga & Yoshino, 1975. Tanabe Bay, Japan.

Subtropical Japan to northern Indonesia and Papua New Guinea. Identified by the three spots in the caudal fin and black-margined narrow blue stripe along upper sides of the body. Easily confused with similar *D. melanopleura* that has a wider blue stripe and different caudal fin markings. Inhabits sheltered rocky reefs, usually in pairs and swims near sponges or long-spined urchins. An active cleaner that shares crevices with shrimps and sometimes moray eels. Reported from tide pools to depths of at least 25 m off shore, usually at moderate depths in subtropical zones, but in the tropics mostly seen in shallow depths. Length to 85 mm.

D. japonicus. Owase, Japan. Length 65 mm.

B

D. japonicus. Kochi Pref, Japan. Brooding male, length 85 mm. Tomonori HIRATA.

C

D. japonicus. Milne Bay, PNG. Length 65 mm. Living with the Spot-face Moray *Gymnothorax fimbriatus.*

D

D. japonicus. Lembeh Strait, Sulawesi, Indonesia. L 45 mm.

E

D. japonicus. Izu Pen, Japan. L. 7 cm. Hiroyuki UCHIYAMA.

Double-chin Pipefish *Doryrhamphus bicarinatus*
Doryrhamphus bicarinatus Dawson, 1981. Sodwana Bay, South Africa.

West Indian Ocean, eastern Africa to the Andaman Sea, and appears to be widespread in the Indian Ocean. Caudal fin markings a large single 'glowing' spot. Males have two bony knobs under the snout. Little known species, reported to 28 m depth. Length to 80 mm.

D. bicarinatus. Sodwana, KwazuluNatal. Length 48 mm. Phil HEEMSTRA.

Fantail Pipefish *Doryrhamphus paulus*
Doryrhamphus paulus Fritzsche, 1980. Sorrocco Isl, Mexico.

Widespread in the eastern Pacific with some geographical variations between mainland waters and oceanic populations. Closely related to *D. melanopleura*. Shallow rocky reefs, but reported to 45 m depth. Length to 85 mm.

A **D. paulus.** Galapagos. L. 50 mm. Paul HUMANN.

B *D. paulus.* Playa Venao. Panama. L. 70 mm. G.R. ALLEN.

Max Gibbs' Pipefish *Doryrhamphus* sp 3

Only known from the photograph and appears to represent a new species, closely related to *D. negrosensis*. Specimens are needed for a description.

D. sp 3. Marquesas Islands, French Polynesia. L 65 mm. Max GIBBS.

Queensland Flagtail Pipefish *Doryrhamphus malus*
Choeroichthys suillus malus Whitley, 1954. Queensland, Australia.

Only known from Queensland Australia. Sheltered inner-reef flats and lagoons and islands inside the Great Barrier Reef. Usually under large pieces of coral rubble in less than 10 m depth. Length to 75 mm.

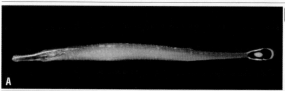

D. malus. Queensland, Australia. **A** Lizard Island. Length 55 mm. Phil Heemstra. **B** One Tree Island, . L. 75 mm.

Flagtail Pipefish *Doryrhamphus negrosensis*

Doryrhamphus negrosensis Herre, 1933. Philippines.

Guam. Depth 3 m. Length 50 mm.
Philippines and Micronesia to northern New Guinea and sporadic records from east Indian Ocean and New Hebrides that
need further investigation. Bluish grey with white stripe over
snout. Sheltered rocky reefs and lagoons. Usually in narrow
crevices with long-spined urchins. Length to 55 mm.

A

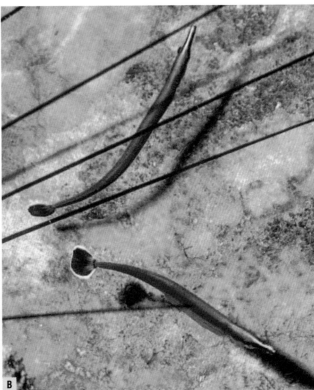

B

D. negrosensis. Guam. Length 50 mm.

Cleaner Pipefish *Doryrhamphus janssi*

Dentirostrum janssi Herald & Randall, 1972. Palau Isl.

Widespread West Pacific, Philippines to northern Australia,
and east Indian Ocean, Christmas Island and Andaman Sea.
Much more elongated and longer snout than other members
in genus. Sheltered inner reefs, usually in caves with sponges
and below large plate corals. Very active cleaner and has
cleaning station that is visited by apogonids and damsels
where adults work in pairs. Usually swims upside-down.
Length to 14 cm.

A

D. janssi. Maumere, Flores, Indonesia. L. 85 mm.

B

D. janssi. Pulau Putri, Java, Indonesia. Largest about 12 cm. Cleaning damsel, *Neopomacentrus anabatoides* (Bleeker).

Masculine. Type species: *Syngnathus dactyliophorus* Bleeker, 1853. Replacement name for *Acanthognathus* Duncker, 1912. Previously treated as subgenus of *Doryrhamphus* by some authors. Comprises at least 8 Indo-West Pacific species, restricted to tropical zones. Free-swimming, usually found inside or in the front of caves or reef overhangs and some are known as being active cleaners. Most species are distinctly barred and have moderately large caudal fins with diagnostic colours. Adults usually occur in pairs. Males incubate the brood below the trunk and eggs are exposed to the outside, partly embedded in the skin. Eggs vary in size and number between 30 to about 200 per brood, depending on the species.

Red-stripe Pipefish *Dunckerocampus baldwini*

Doryrhamphus baldwini Herald and Randall, 1972. Hawaiian Is.

Endemic to the Hawaiian region, where they are known from Maui and Oahu. Unlike most members in the genus, this species lacks vertical bands around the body and has a red stripe along its entire length. The red caudal fin with white upper and lower tip is shared by several other species of *Dunckerocampus* in the Indian Ocean. Usually found in caves or amongst rocks to about 50 m depth, but reported from over 100 m. Males may brood as many as 200 eggs (**A**). Length to 15 cm.

D. baldwini. Eggs on the trunk of the male. Hawaii, Rob Myers.

D. baldwini. Female. Hawaii, Rob Myers.

Christmas Island Pipefish *Dunckerocampus* sp 1

Only known from the photographed specimen. Very similar to *D. baldwini* from Hawaii, sharing the longitudinal stripe, instead of the barred pattern like in most other members of the genus. More specimens are needed. Expected to live in caves like its congeners. The specimen was collected with ichthyocides during a survey of the region.

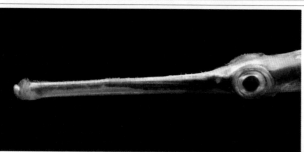

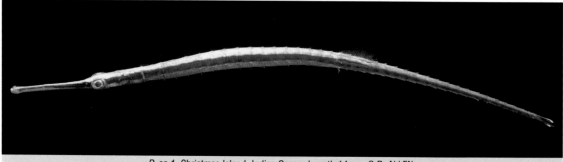

D. sp 1. Christmas Island, Indian Ocean. Length 14 cm. G.R. ALLEN.

Broad-banded Pipefish
Dunckerocampus boylei

Dunckerocampus boylei Kuiter, 1998.
Mauritius.

Widespread Indian Ocean and Red Sea. A little known species, previously confused with the common banded pipefish *D. dactyliophorus* and distinguished from that species by its broader bands and diagnostic caudal fin pattern. Appears to prefer depths of about 25 m or more and is easily overlooked. Only single individuals were observed in the back of caves but the holotype, a male, and a paratype, a female, were collected together. The largest specimen measured just over 16 cm.

D. boylei. Red Sea. Thomas PAULUS.

D. boylei. Red Sea. Thomas PAULUS.

D. boylei. Tulamben, Bali, Indonesia. Length 16 cm. Takamasa TONOZUKA.

D. boylei. Tulamben, Bali, Indonesia. Length 15 cm. Bill BOYLE.

A

D. dactyliophorus. Maumere, Flores, Indonesia. L. 15 cm.

Banded Pipefish *Dunckerocampus dactyliophorus*

Syngnathus dactyliophorus Bleeker, 1853. Java, Indonesia.

West Pacific but numerous records are based on other banded species. Records from Indian Ocean and Red Sea are all erroneous, often illustrated with photographs that were taken in the West Pacific. Caudal fin typically coloured as shown in photographs. Variable from white with a large red ring to mostly red with white margins and centre. Dark bands on the body nearly black in adults and about equal to pale interspaces, including on tail and head. A shallow water species, commonly found inshore and outer reef lagoons, sometimes forming groups with numerous individuals (**A**), but adults usually seen in pairs. Males carry eggs below the trunk (**C**) that are red when fresh. They hatch after a few weeks and young are almost transparent, and pelagic to about 30 mm (**D**). Pigmentation begins when settling on the substrate. Juveniles often near tidal zone and in rockpools. Adults to about 10 m depth. Reports from deep water are based on other banded species. They are often seen in large caves with shrimps or moray eels and engage in cleaning activities, probably picking small copepods or other parasitic crustacea from other fishes. This species is commonly seen in aquaria and is easily kept, ideally in an invertebrate dominated set-up, and provided with a regular supply of small shrimps or mysids. Length to about 18 cm.

This species is often available for the aquarium. They are best kept in pairs and a small cave should be created for a home.

B

D. dactyliophorus. Pulau Putri, Java, Indonesia. Length 16 cm.

C

D. dactyliophorus. Tulamben, Bali, Indonesia. Male with brood. L. 16 cm.

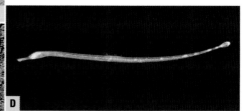

D

D. dactyliophorus. Tomini Bay, Sulawesi, Indonesia. Night, open ocean. Pelagic young, about 30 mm.

E

D. dactyliophorus. Maumere, Flores, Indonesia. Length 10 cm.

F

D. dactyliophorus. Great Barrier Reef, Australia. Length 15 cm. Phil WOODHEAD.

Sulawesi Pipefish
Dunckerocampus sp 2
Undetermined species

Appears to be undescribed and known from northern Sulawesi, Solomon Isl, Micronesia and southern Japan. Occurs at moderate depths in 15–25 m. Very similar to *Dunckerocampus dactyliophorus*, but dark bands broad ventrally, and those posteriorly on tail are wider than pale interspaces. Caudal fin small, and marked like *D. multiannulatus* from the Indian Ocean. Specimens required. Length to about 20 cm.

A

D. sp 2. Lembeh Strait, Sulawesi. Depth 29 m. Juvenile, length 65 mm.

B

D. sp 2. Manado, Sulawesi, Indonesia. Adult, length about 16 cm. Ed ROBINSON.

C

D. sp 2. Guam. Depth 25 m. Brooding male, length about 20 cm. Tim ALLEN.

Glow-tail Pipefish
Dunckerocampus chapmani
Dunckerocampus caulleryi chapmani
Herald, 1953. Noumea, New Caledonia.

Only known from the New Caledonia region. Occurs in shallow lagoon habitats in a few metres depth. A distinctly banded species with a highly recognisable tail. Has large eggs, about pupil-sized, and the low number of about 30 per brood is probably the reason for its highly localised occurrence. The smallest member in the genus, length to about 10 cm.

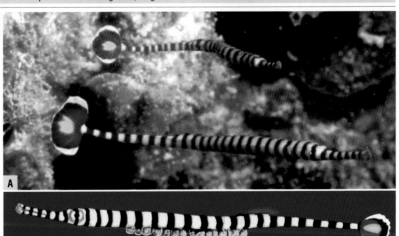

A

B

D. chapmani. Noumea Aquarium. Length 10 cm. **A** pair. Robert Myers. **B** brooding male.

97

Yellow Banded Pipefish
Dunckerocampus pessuliferus

Dunckerocampus pessuliferus Fowler, 1938. Sulu Archipelago.

West Pacific from Philippines to southern Indonesia, along Wallace line, and northern Sulawesi. Records of *D. multiannulatus* from the West Pacific are based on this species. Seems to prefer muddy substrate and usually seen by divers in deep waters that are not disturbed by waves, but it enters shallow estuaries. Adults form pairs and usually swim along the bottom around large remote coral heads on mudslopes. An active cleaner and often swims up to the camera, checking if it is a potential customer. Length to 16 cm.

D. pessuliferus. Tulamben, Bali, Indonesia. Juvenile, 10 cm.

D. pessuliferus. Gilimanuk, Bali, Indonesia. Juvenile, 12 cm.

D. pessuliferus. Monte Bellos Islands, Western Australia. Clay BRYCE.

D. pessuliferus. Lembeh Strait, Sulawesi, Indonesia. Adult, length about 16 cm, and an undescribed species of *Apogon* in front.

Many-bands Pipefish
Dunckerocampus multiannulatus

Doryrhamphus (Dunckerocampus) multiannulatus
Regan, 1903. Mauritius.

Widespread Indian Ocean from east Africa to
Andaman Sea, reaching Sumatra, Indonesia.
Sheltered reefs, usually in the back of caves or
large overhangs to about 20 m depth. Adults
often in pairs. Nearly always seen swimming
upside down against ceiling of caves. Combined
colour patterns of caudal fin and body are diagnostic. Length to 19 cm.

D. multiannulatus. Andaman Sea, Thailand. Mark STRICKLAND.

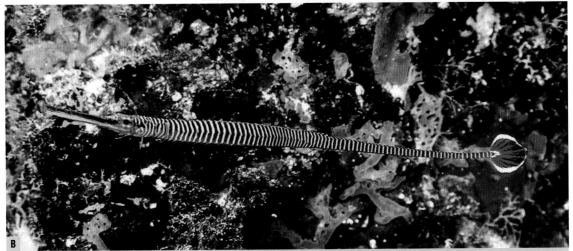

D. multiannulatus. Maldives. Length about 16 cm.

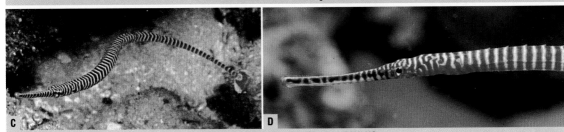

D. multiannulatus. Red Sea. Thomas PAULUS.

D. multiannulatus. **E** underside of male with brood in hatching stage. **F** some of the hatchlings, ~ 5 mm long. Red Sea. Thomas PAULUS.

Feminine. Type species: *Maroubra perserrata* Whitley, 1948. Only 2 species known, divided between sub-tropical Japan and Australia. Males carry brood below trunk and have large eggs, numbering about 60-140. Both species live on rocky reef habitats with invertebrate-rich crevices, sometimes in loose groups.

Orange Pipefish
Maroubra yasudai

Maroubra yasudai Dawson, 1983.
Izu Peninsula, Honshu, Japan.

Only known from the Izu Peninsula area. Found on rocky reefs in moderate depths of about 25 m or more. Swims close to substrate in narrow crevices with rich invertebrate growth. Males brood eggs below trunk (**A-C & E**). Like in other species that carry the brood similarly, the ventral area of the male is conditioned with sperm, at the same time when the female is ready to lay her eggs. The skin becomes soft and after transfer it reacts to the eggs by raising around each, fertilising it and hardening into a short cup to hold it (see close-ups). Length to 15 cm.

C   *M. yasudai*. IOP, Izu Peninsula. Depth 25 m. Length 15 cm. Male with large brood, covering most of trunk, expanding sideways (**B**).

D   *M. yasudai*. Osezaki, Izu Peninsula, Suruga Bay, Japan. Undetermined sex, length 15 cm. Hiroyuki UCHIYAMA.

E   *M. yasudai*. Osezaki, Izu Peninsula, Suruga Bay, Japan. Male with brood, insert close-up of brood. Length 13 cm. Hiroyuki UCHIYAMA.

Sawtooth Pipefish
Maroubra perserrata

Maroubra perserrata Whitley, 1948.
Maroubra, New South Wales, Australia.

Widespread along Australia's south coast from northern New South Wales to southern Western Australia, including Tasmania. Semi-exposed to open ocean rocky shore reef habitat. Found in the back of narrow crevices, often behind urchins, and usually adults occur in pairs, but sometimes form small aggregations. Males incubate brood below trunk with eggs exposed (**A**). They swim close to the substrate, usually resting with tail on the bottom, and feed on tiny crustaceans that crawl on the substrate, such as young amphipods that commonly crawl on sponges. Also feed on zooplankton that drift close to the bottom and through the crevices. Length to 85 mm.

This species was easily kept and readily bred in aquariums in Sydney and Melbourne where small mysids are readily collected in the wild. They live for about 2 years and breed within the first year from birth. They produce a brood about every month for several months in summer time. Eggs are comparatively large and number about 60 on fully grown individuals. Hatching after about 22 days of incubation.

A

M. perserrata. Sydney Harbour. New South Wales, Australia.Male with brood under trunk. L. 65 mm.

B

M. perserrata. Montague Isl, New South Wales, Australia. L. 70 mm.

C

M. perserrata. Seal Rocks, New South Wales, Australia. Length 65-80 mm.

Feminine. Type species: *Heraldia nocturna* Paxton, 1975. Only represented with two sibling species in southern Australian waters with an eastern and western distributions. Secretive in the back of low caves in rocky reefs, normally swimming upside-down on the ceilings. Feeds during the day, not at night, as sometimes reported, but is usually found by divers at night because they look in caves with their torches. Males broods up to about 100 openly exposed eggs below trunk.

A

H. nocturna. Watson's Bay, Sydney Harbour, Australia. L. 70 mm, probably female.

B

H. nocturna. Watson's Bay, Sydney Harbour, Australia. Length 35 mm, juvenile.

C

H. nocturna. Sydney Harbour, Australia. L. 75 mm. Male with brood, about 80 eggs.

Eastern Upside-down Pipefish
Heraldia nocturna

Heraldia nocturna Paxton, 1975.
Watson's Bay, Sydney Harbour, New South Wales, Australia.

Only known from New South Wales, Seal Rocks to Jervis Bay, but may occur further north or south. Sheltered coastal coves and harbours. Lives secretively in rocky reefs to about 20 m depth, and usually only seen at night when shining a light in caves. Typically swims upside-down on the ceiling of caves and adults nearly always found in pairs. Does well in captivity and readily breeds. Swims normal way up when on the bottom, but when swimming from ceiling of its shelter to feed in the open it remains upside-down. Length to 8 cm.

A

H. sp 1. Portsea, Port Phillip Bay, Victoria, Australia. L. 9 cm, sex not known.

B

H. sp 1. Port Phillip Bay, Victoria, Australia. L. 10 cm, male with brood, about 90 eggs.

C

H. sp 1. Castle Rock, south Western Australia. Depth 10 m. Length 75 mm.

Western Upside-down Pipefish
Heraldia sp 1

Only known from Port Phillip Bay, Victoria to Geograph Bay, south Western Australia. Previously included in above species, but now recognised as a valid separate species that differs in colour and meristics. Coastal bays that are well protected from ocean swell. Low reef, in the back of caves, usually swimming upside-down on ceilings. Adults in pairs. Only seen at night or when looking in dark crevices with a light during the day. Length to 10 cm.

Several pairs were kept in captivity for a number of years and they readily reproduced, but no attempts were made to raise the pelagic young.

Presently 42 genera of pipefishes are placed together in this large sub-family, but future studies may separate the specialised groups in their own taxon. The various species range from tiny, worm-like creatures, to large, stiff-bodied species, some with caudal fins and others with thin flexible prehensile tails. The largest genera number about 20–30 species, but there are several with just one or two species. These are primarily distributed in cooler zones of the Atlantic and Mediterranean Seas, American Pacific and China Seas, and appear to represent an older form that gave rise to many genera in the Indo-West Pacific. The greatest species diversity is in tropical and sub-tropical zones of the Indo-West Pacific. Species-numbers decrease rapidly towards temperate zones, but in sub-temperate waters diversity is high on a generic level, and species are often localised. Tropical genera are usually represented with more species, which are divided between various areas of a large geographical zone. Tropical species are generally small, ranging from a few centimetres to about 40 cm, but most species are less than 20 cm long. The subtropical and sub-temperate genera, being less specious and often monotypic, have probably resulted from specialisation-demands and changing environments since the glacial episodes. They grow largest in sub-temperate zones where some reach 65 cm.

Except for some very habitat specific species, pipefishes are easily kept in captivity and most will breed, especially the small species. They require lots of space in relation to their size, and surroundings that are natural, similar to their general habitat, and breeding only happens when kept in the correct temperatures, similar to from where specimens originated. Diet comprises small crustaceans, and ideally they are fed on a regular basis with live food such as mysids and copepods. Most hatchlings are easily raised as they are born relatively large and advanced.

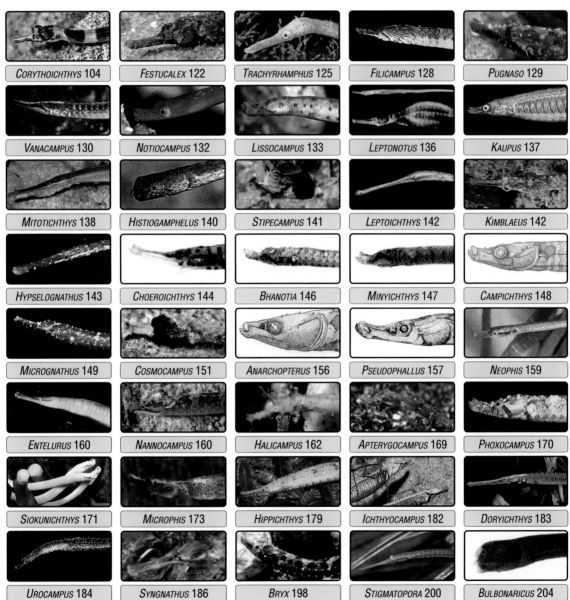

CORYTHOICHTHYS 104	*FESTUCALEX* 122	*TRACHYRHAMPHUS* 125	*FILICAMPUS* 128	*PUGNASO* 129
VANACAMPUS 130	*NOTIOCAMPUS* 132	*LISSOCAMPUS* 133	*LEPTONOTUS* 136	*KAUPUS* 137
MITOTICHTHYS 138	*HISTIOGAMPHELUS* 140	*STIPECAMPUS* 141	*LEPTOICHTHYS* 142	*KIMBLAEUS* 142
HYPSELOGNATHUS 143	*CHOEROICHTHYS* 144	*BHANOTIA* 146	*MINYICHTHYS* 147	*CAMPICHTHYS* 148
MICROGNATHUS 149	*COSMOCAMPUS* 151	*ANARCHOPTERUS* 156	*PSEUDOPHALLUS* 157	*NEOPHIS* 159
ENTELURUS 160	*NANNOCAMPUS* 160	*HALICAMPUS* 162	*APTERYGOCAMPUS* 169	*PHOXOCAMPUS* 170
SIOKUNICHTHYS 171	*MICROPHIS* 173	*HIPPICHTHYS* 179	*ICHTHYOCAMPUS* 182	*DORYICHTHYS* 183
UROCAMPUS 184	*SYNGNATHUS* 186	*BRYX* 198	*STIGMATOPORA* 200	*BULBONARICUS* 204

Masculine. Type species *Syngnathus haematopterus* Bleeker, 1851. Tropical, Indo-Pacific genus with at least 23 species variously distributed in the area. This includes members of species-complexes that were previously treated as a single widespread taxon. Because the present level of taxonomy of syngnathids is primarily based on morphology, most species can't be distinguished this way. Many species in this genus are undescribed.

SPAWNING SEQUENCE P110

Members of *Corythoichthys* are primarily found on rubble zones of reefs and surrounding corals or rocks. They usually occur in pairs during the day, but on open rubble may form small aggregations. Some species congregate at night in large numbers, seeking safe shelter on large sponges that may irritate predators such as moray eels that hunt at night. Males have a large brood pouch under tail, accommodating several hundreds to about 1000 eggs, depending on species and age. After extensive courtship, the eggs are laid as a rolled sheet that unfolds against the body of the male, and a thin layer of skin raised along both sides of the brood, meeting along the centre and usually covering all the eggs, forming a simple pouch. Small near transparent hatchlings appear after a relatively short incubation period of a few weeks and swim to the surface. They may travel a fairly long distance as pelagic larvae, but few species are widespread, and much more localised than previously thought. Colour-variations are usually related to habitat, stage of development, or sex. Geographical variation in some of the species that are more widespread have localised forms that may represent subspecies. In species-complexes, representatives that are geographically far apart can look very similar, whilst those closest together are the most different. Underwater, the species can be identified by a combination of features, usually in combination of length of snout and colouration of the head and body. Male and female may differ in colouration of the head or chest, and general colouration varies between different habitats, but certain patterns are diagnostic. Diet comprises tiny crustaceans that are usually picked from the substrate. All species are easily kept in captivity, providing that their specialised and regular feeding requirements are met.

C. *amplexus*. Santo, Vanuatu, 30 m. Length 75 mm. Neville COLEMAN.

Fijian Banded Pipefish *Corythoichthys amplexus*

Corythoichthys amplexus Dawson & Randall, 1975. Fiji I.

Reported throughout the Indo-West Pacific, but probably restricted to the Fijian region. Pale interspaces grey with white edges. A short-snouted and broadly banded species with several similar species elsewhere. Lives in moderately deep water on reefs or along the edges on rubble. Length to 85 mm.

Broken-bands Pipefish *Corythoichthys* sp 1

Only known from the Great Barrier Reef, Australia. Previously included with *C. amplexus*, but requires further investigation. Snout not as short. Dark bands broken-up into spots. Head spotted all over (**C**) or blotched (**D**). Caudal fin all white. Usually shallow on reef crests from 5 to 20 m. Adults in pairs and they congregate in large groups at night on sponges in their 'safety house'. Appears to be a small species, length to 85 mm.

A

C. sp 1. Lizard I, Qld. Female, 75 mm. Neville COLEMAN.

C

C. sp 1. Heron I, Qld. Neville COLEMAN.

B

C. sp 1. Heron I, Qld. Brooding male, 85 mm. Neville COLEMAN.

D

C. sp 1? Lizard I, Qld, Australia 75 mm.

Red-spot Pipefish
Corythoichthys sp 2

Known from Bali & Flores, Indonesia, and PNG to northern Great Barrier Reef. A short-snouted species that appears to be undescribed. Closely related to *C. amplexus* but dark bands narrow or less distinct and it grows much larger. Each dark band followed by a narrower white one, snout with small red spots, and female with a distinct bright red spot ventrally on the first dark body band (**A**). On mixed coral and algae reef to about 25 m depth. Length to 12 cm.

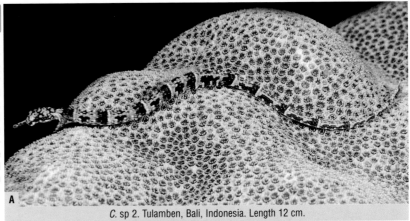

A

C. sp 2. Tulamben, Bali, Indonesia. Length 12 cm.

B

C. sp 2. Great Barrier Reef, Australia. Length 10 cm. Phil WOODHEAD.

C

C. sp 2. Maumere, Flores, Indonesia. Length 12 cm.

D

C. sp 2. Maumere, Flores, Indonesia. Length 12 cm.

E

C. sp 2. Madang, PNG. L. 10 cm. Neville COLEMAN.

F

C. sp 2. Madang, PNG. L. 10 cm. Neville COLEMAN.

A

C. sp 3. Maumere, Flores, Indonesia. Female, 85 mm.

West Pacific from southern Japan to southern Great Barrier Reef, Australia. Previously included with *C. amplexus.* Snout very short. Dark bands much broader than pale inter-spaces. Head in front of eyes mostly white, snout with dark lateral stripe or spots. Secretive on sheltered coral reefs, mainly inshore, and to depths of about 20 m. Length to 90 mm.

Remarks: This species is generally reported or referred to as *C. amplexus.*

B

C. sp 3. Northern Great Barrier Reef, Australia. Brooding male, 85 mm. Phil WOODHEAD.

C

C. sp 3. Northern Great Barrier Reef, Australia. Female, 75 mm. Phil WOODHEAD.

D

C. sp 3. One Tree I, Qld, Australia 60 mm.

E

C. sp 3. Milne Bay, PNG 65 mm.

F

C. sp 3. Maumere, Flores, Indonesia. Brooding male, 90 mm.

Oman Pipefish *Corythoichthys* sp 4

Undetermined species from Oman. Short-snouted and may have been reported from the areas as either *C. amplexus* or *C. flavofasciatus*. Appears to be similar to several other species in the West Pacific and highlights the problems when working with living colours versus morphology or when working with preserved material only. Needs further investigation. Length to about 10 cm.

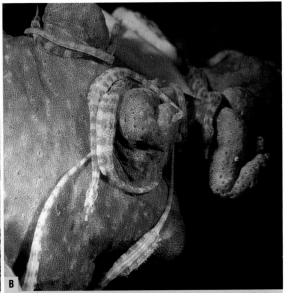

C. sp 4. Oman. Length 10 cm. Phil WOODHEAD.

C. sp 4. Oman. Length 10 cm. Night gathering. Phil WOODHEAD.

Red-scribbled Pipefish
Corythoichthys sp 5

West Indian Ocean. A short-snouted species that is usually reported as *C. amplexus*. Appears to be undescribed and distribution is unknown, probably restricted to the western Indian Ocean. Possibly identical to above species from Oman and seems closely related to *C. amplexus* and sp 1. Needs further investigation.

C. sp 5. Seychelles. Length 9 cm. Neville COLEMAN.

Yellow-red Pipefish
Corythoichthys sp 6

West Indian Ocean. Appears to be undescribed and its distribution is unknown, probably is restricted to the western Indian Ocean. Its colouration is distinctive and it lacks the reticulations of other similar species. A short-snouted species that is usually reported as *C. amplexus* or *C. flavofasciatus*. Banding is in various colours, changing from dusky blotches anteriorly to bright red posteriorly with yellow lines or blotches in the pale areas. It is unlike *C. amplexus* that appears to be restricted to the West Pacific, but the head pattern and snout length suggest a close relationship with *C. flavofasciatus*. Needs further investigation.

C. sp 6. Seychelles. Length 11 cm. Roger. C. STEENE.

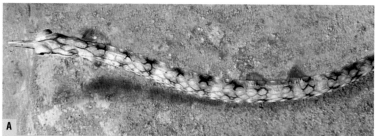

A

Corythroichthys conspicillatus
Jenyns, 1842. Tahiti.

Coral Sea to Central Pacific. Body with pattern of black and yellow reticulations, forming an indistinct pattern of alternating dark and greenish bands. Previously included with *C. flavofasciatus,* a short-snouted species from the Red Sea. Length to 18 cm.

C. conspicillatus. Yonge reef, Great Barrier Reef. Female, length 15 cm. Neville COLEMAN.

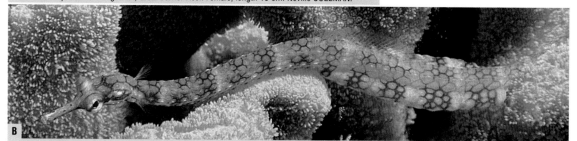

B

C. conspicillatus. Guam. Brooding male, length 18 cm. Rob MYERS.

A

Yellow-banded Pipefish
Corythoichthys flavofasciatus

Syngnathus flavofasciatus Rüppell, 1838. Jeddah, Red Sea.

Red Sea and Indian Ocean. A short-snouted species, usually with an irregularly banded pattern that may be broken up in large blotches and some reticulations on the trunk. Yellow streaks or bands in the pale interspaces. Mainly seen on rubble reef, usually in pairs, to depths of 25 m. Records from elsewhere are based on other species. Length to 18 cm.

C. flavofasciatus. Egypt, Red Sea. Length. 12 cm.

B

C. flavofasciatus. Egypt, Red Sea. Length 18 cm.

C

C. flavofasciatus. Maldives. Length 18 cm.

108

Ishigaki Pipefish
Corythoichthys isigakius

Corythroichthys isigakius
Jordan & Snyder, 1901. Ryu Kyu I, Japan.

Probably restricted to tropical and sub-tropical waters of Japan. Rocky and coral reefs in depths of about 10–15 m. Previously included with *C. haematopterus*, but whilst the morphology maybe very similar, the living colours show little resemblance between the two species, but is rather more similar to *C. conspicillatus* (see opposite page). Length to 18 cm.

SPAWNING SEQUENCE P110

A

C. isigakius. Izu Peninsula, Japan. Length 10 cm. Hiroyuki UCHIYAMA.

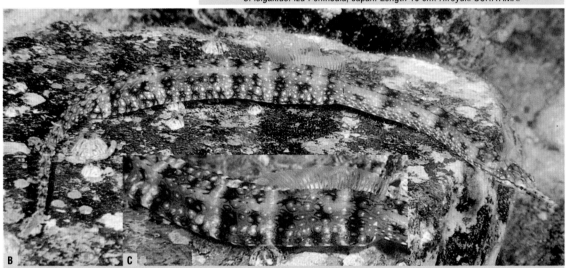

B   C

C. isigakius. Kashiwajima, Kochi Japan. Brooding male, length 18 cm. **C** close-up of pouch.

D

C. isigakius. Kashiwajima, Kochi Japan. Gravid female, length 18 cm.

Spawning behaviour. Ehime Pref, Japan. Depth 6 m, length 15 cm.

Photograph sequence by Tomonori HIRATA

Like many pipefishes, most members in the genus are strongly pair-bonded and usually spawn with the same partner repeatedly. This kind of relationship is often termed *Monogamy*, which in principal means pairing for breeding purposes, but this term is often hyped into a life-part-nership, and latter is not true for any know fish species. Many species will stay together for a season but may change partners. I prefer using pair-bonding instead of *Monogamy* to avoid misinterpretation.

Courtship and spawning behaviour of several species of *Corythoichthys* has been observed and filmed on the Great Barrier Reef in Australia, and is here presented with a sequence of superb stills of the Japanese **Ishigaki Pipefish**. Once these fishes form pairs, they greet each other every morning on day break by circling each other with head raising and turning the trunk to show certain markings below. This courtship rou-tine intensifies as the female becomes obviously gravid and the male is nearing his readiness to incubate the brood. They may go through a number of practice spawning runs for several days, so when the time comes, they know each others moves. In the pictures below (**A–C**) they are preparing for actual spawning, which begins by repeatedly swimming up from the bottom, drifting along for a few seconds, and back to the bottom, until both are ready. The female with her greatly swollen trunk is ready as her prepared ovipositor shows. In **B** the flattened tail of the male shows his readiness to receive the brood. Together they rise from the substrate where they manoeuvre themselves into position. The eggs are suddenly produced (**D**) as the female swims upwards against the male, in so doing the eggs are put on from the tail-end towards the trunk (**E**). Then she assist in securing the eggs to the body of the male by pushing the egg mass with her body into place. Swimming just above the bottom, the twisting motions of their bodies looks if they are dancing along.

After a few weeks the babies hatch. Often the break the egg shell and their tails can be seen hanging out. When they are about ready to come out completely, the male begins to jerk his body sideways that may dislodge some of the babies. Then he swims up just a little from the bottom and may violently shake all of them out in a few seconds. They rapidly swim towards the surface to begin their pelagic stage.

Note: some photographs were slightly rotated more vertically to fit in the layout.

BACK TO
GENUS INTRO P. 104
SPECIES ACCOUNT P.109

A   B   C

Undetermined species that may have been confused with *C. haematopterus* or *C. flavofasciatus*. Series of black spots along back ridges and reticulations along sides. A net work of yellow lines over most of the body. Caudal fin pinkish. Silty inshore reefs. Length to 18 cm.

C. sp 8. Pulau Putri, Java, Indonesia. Brooding male, length 18 cm.

C. sp 8. Pulau Putri, Java, Indonesia. Length 18 cm.

C. sp 8. Karimun Is, Java, Indonesia. Depth 5 m. Length 18 cm. Takamasa TONOZUKA.

Reef-top Pipefish
Corythoichthys haematopterus

Syngnathus haematopterus Bleeker, 1851.
Banda Neira, Indonesia.

Widespread West Pacific. Sheltered inner reef flats and rubble lagoons, usually semi-silty zones. Intertidal and generally shallow, to about 10 meters depth. Adults nearly always in pairs. Fine lines on the head and series of spots over the back and along trunk. Caudal fin mainly pink. Length to 18 cm.

A

C. haematopterus. Maumere, Flores, Indonesia. Length 18 cm.

B

C. haematopterus. Singapore. Length 16 cm.

C

C. haematopterus. Maumere, Flores, Indonesia. Length 18 cm.

Maldives Pipefish
Corythoichthys sp 9

Possibly restricted to Maldives and Andaman Sea. Reported as *C. haematopterus* that is probably only in the West Pacific. A photograph of this species, taken in the Maldives, was used in J. E. Randall's recent book 'Coastal Fishes of Oman'. Head markings are distinctive and lacks the typical fine-lined pattern on the head of *C. haematopterus* (see **B & C**), as well as the fine spotting on the ventral trunk area and having a generally striated pattern along entire body. Sheltered rubble zones in lagoons and harbours, in pairs or small aggregations, usually in a few metres depth. Length to about 18 cm.

A

C. sp 9. Maldives. Length 18 cm.

B

Close-up from **C**, typical head-pattern.

C

Close-up of *C. haematopterus*, typical head-pattern.

D

C. sp 9. Maldives. Pair, brooding male above, length 18 cm.

Banded Messmate Pipefish
Corythoichthys sp 10

Known from Flores and Moluccan Seas in Indonesia, but also found recently at the Rowley Shoals off western Australia. Reported as *Corythoichthys flavofasciatus* and *C. haematopterus* in some books. Generally has a more striated pattern compared to similar species, lacking any vermiculation of the lines, and even alternating dark and yellow banding along the entire body. A somewhat stocky species compared to the others mentioned. Usually found on algae and coral reefs to about 15 m depth or nearby on coarse sand and rubble. Often in small aggregations under jetties that are typically built in protected areas, usually in large lagoons. Caudal fin pinkish with pale margin like in many other species in the genus that venture onto rubble and sand along reef margins. Length to 17 cm.

C. sp 10. Maumere, Flores, Indonesia. Length 17 cm.

C. sp 10. Lembeh Strait, Sulawesi, Indonesia. Length 17 cm.

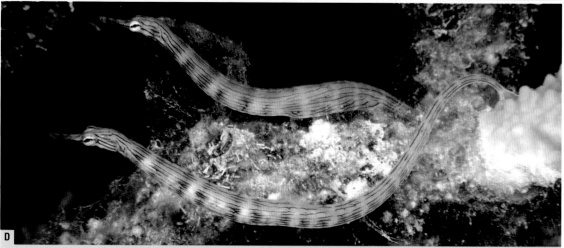

C. sp 10. Rowley Shoals, off north-western Australia. Depth 10 m. Length 16 cm.

Pacific Messmate Pipefish
Corythoichthys waitei

Corythroichthys waitei Jordan & Seale, 1906. Samoa.

West Pacific, Philippines, Micronesia to Samoa. Previously included with *C. intestinalis* but has a longer snout which is strongly marked with black lines. Head and snout with distinct lines and body with a striated pattern, mixed with reticulations. Indistinct and uneven banding along the body. Sheltered in shallow lagoons and harbours, usually in few metres depth. During the day in pairs or forms small aggregations at night. Length to about 18 cm.

C. *waitei*. Guam. Length 16 cm.

C. *waitei*. Guam. Length 16 cm. Rob MYERS.

Egyptian Pipefish
Corythoichthys sp 11

Probably a Red Sea endemic. A long-snouted species that has been confused with *C. schultzi*. Has a black line through the eye, and series of pupil-sized white spots on trunk, one mid-laterally and other along lower ridge. Males have distinctive white barring over the brooding area. Sheltered bays and harbours, along reef edges on sand. Adults in pairs or small groups, usually in 3–10 m depth. Length to 18 cm.

C. sp 11. Egypt, Red Sea. Female, length 17 cm.

C. sp 11. Egypt, Red Sea. Brooding male, length 18 cm.

A

C. schultzi. Maumere, Flores, Indonesia. Length 17 cm.

Corythoichthys schultzi Herald, 1953. Marshall I.

Widespread Indo-West Pacific from southern Japan to northern Australia and west to African coast. Usually on coarse sand and rubble along edges of reefs. Adults in pairs or small aggregations when in the open or in safe places at night. Indistinctly banded, dark bands followed by narrow pale one. Back ridges with series of orange dashes and sides with elongates spots, sometimes forming series of short lines. Length to 16 cm.

B

C. schultzi. Great Barrier Reef, Australia. Length 17 cm. Phil WOODHEAD.

C

C. schultzi. Kerama, Japan. Length 17 cm.

D

C. schultzi. Maumere, Flores, Indonesia. Length 17 cm.

C. schultzi. Maldives. Brooding male, length 14 cm.

C. schultzi. Maldives. Female, length 14 cm.

Orange-spotted Pipefish
Corythoichthys ocellatus

Corythoichthys ocellatus Herald, 1953. Solomon I.

West Pacific, from Philippines to north-eastern Australia and Solomon I. Very similar to *C. schultzi* but has fewer dorsal fin rays and tail rings. The body is primarily covered with small to pupil-sized orange spots that are often dark-ringed as ocelli, especially over the back of the trunk. Inshore, on algae and rubble reefs to about 15 m depth. Solitary or in pairs. Length to 12 cm.

C. ocellatus. Lizard Island, Qld, Australia. Brooding male, length 12 cm.

C. ocellatus. Off Port Moresby, Papua New Guinea. Neville COLEMAN.

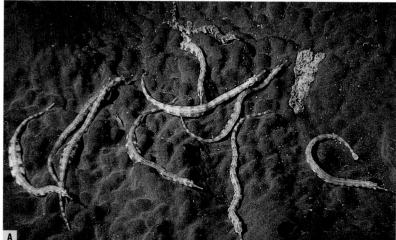

Australian Messmate Pipefish
Corythoichthys intestinalis

Syngnathus intestinalis Ramsay, 1881.
Solomon I.

Southern West Pacific, north-eastern Australia and Papua New Guinea to Solomon Islands. Similar species in northern West and central Pacific. Head with distinct lines and body with a faint banded pattern, mixed with distinct evenly spaced series of black reticulations along entire body and tail. Sheltered sponge reefs, in shallow lagoons and harbours, usually in 5–10 m depth. During the day in pairs or small aggregations, grouping together at night is safe places (**A**). Length to about 18 cm.

C. intestinalis. Grouping together at night. Length to 18 cm. Madang, PNG. Neville COLEMAN.

C. intestinalis. Length 12 cm. Tryon Island, Great Barrier Reef, Queensland, Australia. Neville COLEMAN.

Paxton's Pipefish
Corythoichthys paxtoni

Corythoichthys paxtoni Dawson, 1977.
One Tree I, Great Barrier Reef, Australia.

Only known from the Great Barrier Reef. A yellowish species with series of dark scribbles along most of the body, distinct markings usually fading on the tail. Shallow and protected reefs and rubble lagoons, from intertidal to about 15 m depth. Length to 15 cm.

C. paxtoni. One Tree I, Great Barrier Reef, length 14 cm. Mark NORMAN.

Yellow-spotted Pipefish
Corythoichthys polynotatus

Corythoichthys polynotatus
Dawson, 1977. Palau I.

West Pacific, Palau and Philippines to southern Indonesia. Mainly found in shallow rubble lagoons with algae and seagrasses. Often intertidal, and usually in only a few metres depth. Lacks black lines and primarily a yellow-spotted pattern dorsally. Caudal fin mainly pink. Length to 16 cm.

C. polynotatus. Sanur, Bali, Indonesia. Length. 16 cm.

C. polynotatus. Singapore. Length. 16 cm.

C. polynotatus. Singapore. Length. 16 cm.

C. polynotatus. Maumere, Flores, Indonesia. Brooding make, length. 18 cm.

Corythoichthys insularis
Dawson, 1977. Amiranti I.

West Indian Ocean and Maldives. Distinctly marked on the head and body variable but usually shows three thin white bands on the trunk and males with three on the brooding area. A small species that appears to prefer clear outer reef habitats, usually on rubble with rich invertebrate growth on the floor of large caves. In depths between 15–40 m. Length to 12 cm.

C. insularis. Maldives. Pair, length 12 cm.

C. insularis. Maldives. Brooding male, length 12 cm.

C. insularis. Maldives. Female, length 12 cm.

C. insularis. Maldives. Grouping in a cave. Neville COLEMAN.

Black-breasted Pipefish
Corythoichthys nigripectus

Corythoichthys nigripectus
Herald, 1953. Bikini Atoll.

Micronesia to central Pacific. Only known from oceanic locations. Photographs depicting this species are rare and its illustrations in books are usually substituted with those of sp 12 from the Red Sea, a species that is often seen in popular dive sites. Specimens are known 5 to about 30 m, living on coral reefs. Adults usually in pairs. Length to 11 cm.

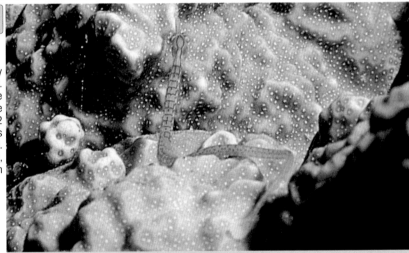

C. nigripectus. Guam. Rob MYERS.

Red Sea Pipefish
Corythoichthys sp 12

Red Sea endemic that has been confused with its sibling *C. nigripectus* from the West Pacific, because of similar meristic values. Occurs on mixed algae and coral reefs, usually clear water habitats and often on rubble patches on the floor of large caves to depths of about 30 m. Adults usually in pairs. Length to 12 cm.

C. sp 7. Egypt, Red Sea. Length 12 cm.

121

GENUS *FESTUCALEX* Whitley, 1931

Masculine. Type species *Syngnathus cinctus* Ramsay, 1882. Small Indo-Pacific genus with about 12 species variously distributed in the area. Nine species are described and several more are only known from juvenile or pelagic stages. Many species are known from few specimens only and all appear to have a restricted distribution and are endemic to relatively small areas.

Generally small, 10–15 cm fully grown, and secretive or highly camouflaged species that inhabit low algae-rubble substrates from moderate depths of tidal channels in estuaries to deep offshore on muddy soft-bottom environment. Most species have never been observed by diving.

A

Girdled Pipefish *Festucalex cinctus*

Syngnathus cinctus Ramsay, 1882.
Sydney Harbour, Australia.

Only known from eastern Australia, probably restricted from central New South Wales to central Queensland. Highly variable in colour but usually the opercle is orange. Found on rubble substrate with short algaes and sponges in 10–20 m depth. Length to 16 cm.

F. cinctus. Sydney Harbour, Australia. Length 12 cm.

B

F. cinctus. Sydney Harbour, Australia. Length 12 cm. Rare white form.

C

F. cinctus. Sydney Harbour, Australia. Length 12 cm. Common colouration.

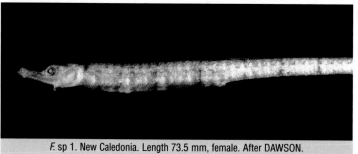

Caledonian Pipefish *Festucalex* sp 1

Uncertain species, probably restricted to the Coral Sea. Previously reported as *F. erythraeus,* but appears to be a variation of *F. cinctus* (above).

F. sp 1. New Caledonia. Length 73.5 mm, female. After DAWSON.

Ladder Pipefish
Festucalex scalaris

Ichthyocampus scalaris Günther, 1870. Shark's Bay, W.A.

Only known from tropical western Australian coast. Intertidal to about 20 m depth. Usually in algae or sargassum weeds.

Resembles *Vanacampus* from temperate southern Australian waters. Length to 20 cm.

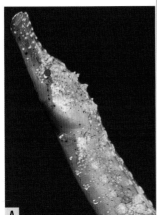

A

B

F. scalaris. **A** close-up, **B** Pair, including brooding male. Western Australia. Depth 3 m. Neville COLEMAN.

Gibb's Pipefish
Festucalex gibbsi

Festucalex gibbsi Dawson, 1977. Queensland, Australia

Known from a few specimens from Irian Jaya to Queensland, Australia, but maybe identical to *F. wassi* from Samoa (below) and Fiji, and also Irian Jaya. The specimen in the photograph is virtually the same as the holotype from Samoa. Most specimens are known from trawls between 20 and 91 m depth. Open rubble and soft-bottom habitat. Length to 10 cm.

F. gibbsi. Lizard Island, Queensland, Australia. Depth 15 m. Neville COLEMAN.

Wass' Pipefish
Festucalex wassi

Festucalex wassi Dawson, 1977. Samoa. Probably a synonym of *F. gibbsi* (above).

F. wassi. Samoa. Brooding male, 69.5 mm SL.. After DAWSON.

F. amakusensis. Yellow variety. IOP, Japan. Hajime MASUDA.

Amakusa Pipefish
Festucalex amakusensis
Hippichthys amakusensis Tomiyama, 1972.
Aitsu, Japan.

Only known from sub-tropical Japanese waters. Variable from dull brown to bright red. Previously included with *F. erythraeus* a Hawaiian endemic. Lives on rocky reef and rubble or coarse sand substrates along reef margins. Length to 16 cm.

F. amakusensis. Osezaki, Japan. Hiroyuki UCHIYAMA.

F. amakusensis. Red variety. IOP, Japan. Hajime MASUDA.

Townsend's Pipefish
Festucalex townsendi
Ichthyocampus townsendi Duncker, 1915.
Makran coast & Maldives.

Western Indian Ocean. Similar to *F. erythraeus*, but fewer tail rings. Reddish in colour. Reported as inshore to 80 m depth and a length of 8 cm.

F. townsendi. Maldives. Syntype, 48.5 mm. After DAWSON.

List of additional described species of *Festucalex*

Burgundy Pipefish *Festucalex erythraeus.* As *Ichthyocampus erythraeus* Gilbert, 1905. Hawaii.
Oceanic Pipefish *Festucalex philippinus.* As *Ichthyocampus philippinus* Fowler, 1938. Sulu Archipelago, and *Festucalex prolixus* Dawson, 1984. Sulu Sea.

Masculine. Type species *Syngnathus serratus* Temminck & Schlegel, 1850. Indo-West Pacific with 3 species that are widely distributed for a pipefish but some geographical variations that need further investigation. The presence of juveniles expatriating to sub-tropical zones during warmer summer months indicate a long pelagic stage that would explain their large geographical range

Planktonic stages have dermal processes (see photograph) that seems to be for floating purposes and specimens over 10 cm long have been collected far offshore. However, species between Pacific and Indian Oceans show considerable differences in colour and some slight drift in meristic values that suggests diversion within all the species as presently defined. Adults occur on semi-open substrates, often in current prone areas of deep lagoons and tidal channels in estuaries.

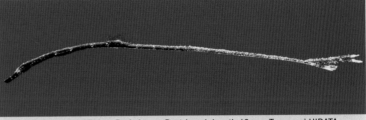

T. bicoarctatus? Ehime Pref, Japan. Post-larval, length 12 cm. Tomonori HIRATA.

Goggle-eye Pipefish
Trachyrhamphus serratus

Syngnathus serratus
Temminck & Schlegel, 1850. Japan.

Sub-tropical waters of Japan and China Seas. Also reported from waters of the Andaman Sea region that probably represents another short-snouted species that was named by Bleeker in 1853 as *Syngnathus subbooko*. Planktonic juveniles stage has dermal processes, dorsally at the tip of the tail (see photo above), that apparently assist in floating. Juveniles have a relatively long snout that proportionally shortens in total length with age and has filaments above the eyes (**A**), similarly to *Filicampus tigris*, to which it seems closely related. Open sand or mud substrates. Reported depth range of 15 to 90 m depth. Length to at least 32 cm.

T. serratus. Izu Peninsula, Japan. Length 15 cm. Hiroyuki UCHIYAMA.

T. serratus. Izu Peninsula, Japan. Length 20 cm. Hiroyuki UCHIYAMA.

T. serratus. Osezaki, Izu Peninsula, Japan. Depth 25 m. Female, length 30 cm. **C** close-up of **D**.

Syngnathus bicoarctatus Bleeker, 1857
Ambon, Indonesia.

Widespread Indo-West Pacific, but probably comprises several similar species that are distributed over the area. The east Australian form looks different and was described as *Yozia compitalis* by Whitley in 1950, from Sydney Harbour, and maybe valid as *Trachyrhamphus compitalis*. All populations need to be investigated further. The Red Sea and Indian Ocean populations are different in shape of the head and have different markings, and those in oceanic locations seem different from those in estuaries. Some populations inhabit seagrass beds and others only rubble sand areas. Most are seen on sand and mud areas, prone to currents. Red Sea population in sheltered bays with seagrasses in few metres depth. Elsewhere usually soft bottom to about 25 m. Length to 40 cm.

T. bicoarctatus. Sydney Harbour, Australia. Length 40 cm.

T. bicoarctatus. Sydney Harbour, Australia. Length 30 cm.

T. bicoarctatus. Milne Bay, PNG. Length 40 cm.

T. bicoarctatus. Flores, Indonesia. Length 30 cm.

T. bicoarctatus. Male. Aliwal Shoal, South Africa. Valda Fraser.

T. bicoarctatus. Egypt, Red Sea. Length 40 cm.

T. bicoarctatus. Maldives. Length 40 cm.

T. bicoarctatus. Pulau Putri, Java. Length 40 cm.

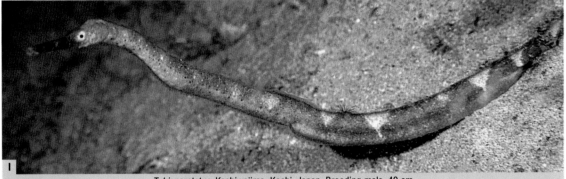

T. bicoarctatus. Kashiwajima, Kochi, Japan. Brooding male, 40 cm.

Straight Stick Pipefish
Trachyrhamphus longirostris

Trachyrhamphus longirostris
Kaup, 1856. Asia.

Reported from throughout the tropical Indo-West Pacific but various populations may represent additional species. Type locality is vague. Indian Ocean form was described as *Syngnathus ceylonensis* Günther, 1870, from Ceylon and Zanzibar. May be confused with the similar *T. bicoarctatus* from which it is distinguished by the lack of appendages on the trunk in the juveniles, head at no angle to the body, and colour patterns. It is less common and is mainly known from deep trawls over muddy substrates, but enters sheltered muddy estuaries where, out in the open, it lays on the bottom. Length to 40 cm.

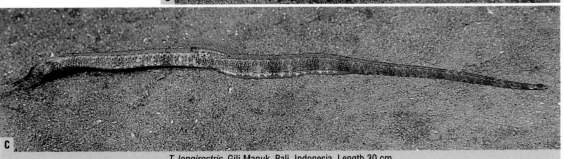

T. longirostris. Gili Manuk, Bali, Indonesia. Length 30 cm.

Masculine. Type species *Sygnathus superciliaris* Günther, 1880 (=*S. tigris* Castelnau, 1879). Monotypic, sub-tropical Australian genus. Related to the tropical Indo-West Pacific genus *Trachyrhamphus*, probably representing their ancestral form closely.

Tiger Pipefish *Filicampus tigris*

Syngnathus tigris Castelnau, 1879.
Sydney Harbour, Australia.

Sub-tropical waters on the east and west coasts of Australia. Old records from the upper Spencer Gulf of South Australia suggest a remnant sub-tropical fauna there, but there are no recent sightings. The area is rather small and vulnerable to pollution. The loss of mangroves and heavy industry may have ended this small population. They are common in Sydney Harbour, where in sheltered bays on sand and muddy substrate adjacent to tidal channels, feeding on the massing mysids. Often in shallow depths along the edges of seagrass beds to at least 30 m depth. Length to 30 cm.

A

B

F. tigris. Sydney Harbour, Australia. Length 30 cm. **B** juvenile stage, about 10 cm.

C

]*F. tigris.* Sydney Harbour, Australia. Brooding male, length 30 cm.

D

F. tigris. Sydney Harbour, Australia. Probably a pair, but difficult to identify a non-brooding male, length 30 cm.

Masculine. Type species *Syngnathus curtirostris* Castelnau, 1872. Sub-temperate Australian genus, closely related to *Vanacampus*. with a single species. Occurs in sheltered bays and estuaries, associating with algae-reef habitats and sometimes *Amphibolis* seagrass beds. Males have a tail pouch with thick skin that overlaps and folds along ventral centre. The flaps interlock that completely protects the brood. Small crustaceans are hunted on the substrate and mysids are an important part of their diet, at least at some juvenile stages.

Pugnose Pipefish
Pugnaso curtirostris
Syngnathus curtirostris Castelnau, 1872.
St. Vincent's Gulf, South Australia.

Australia's south coast from south Western Australia to the Bass Strait region. Occurs on algae reef, often large rubble on sand, and in broad-leaved seagrass species such as *Poseidonia* and *Amphibolis* spp growing amongst low reef. Juveniles often amongst decaying leaf-litter. Sheltered estuaries and bays to about 10 m depth. Length to 15 cm.

P. curtirostris. Portsea, Victoria, Australia. Female, length 11 cm.

P. curtirostris. Flinders, Victoria, Australia. Juvenile, length 6 cm.

V. curtirostris. Juv, Sorrento, Vic.

P. curtirostris. Rye, Victoria, Australia. Length 10 cm.

P. curtirostris. Portsea, Victoria, Australia. Brooding male, length 15 cm.

P. curtirostris. Rye, Victoria, Australia. Female, length 14 cm.

Masculine. Type species *Syngnathus vercoi* Waite & Hale, 1921. Sub-temperate Australian genus with 4 species that are restricted to southern waters, one of which ranging to sub-tropical zones. They are mostly found in sheltered bays and estuaries, associating with algae-reef habitats and some are found in seagrass beds, especially when juvenile. Males have a tail pouch with thick skin that has overlapping folds along ventral centre. The flaps interlock and completely protects the brood. Small crustaceans are hunted on the substrate and mysids are an important part of their diet, at least at some juvenile stages.

Mother-of-Pearl Pipefish
Vanacampus margaritifer

Syngnathus margaritifer Peters, 1869.
Sydney, Australia.

Widespread in southern Australian waters, ranging into cooler coastal waters of southern Queensland. Slight variations in colour between various areas. Occurs on rubble-algae reefs in harbours and estuaries. Often on muddy substrates. To depths of about 10 m. Length usually to 16 cm, but grows larger in southern waters, sometimes to 20 cm.

V. margaritifer. Sydney Harbour, Australia. Courting pair. Length 16 cm.

V. margaritifer. Sydney Harbour, Australia. Length 16 cm.

V. margaritifer. Sydney Harbour, Australia. Length 12 cm.

V. margaritifer. Portsea, Victoria, Australia. Brooding male, length 20 cm.

Port Phillip Pipefish
Vanacampus philippi

Syngnathus philippi Lucas, 1891.
Port Phillip, Victoria, Australia.

Several separate populations along Australia's south coast. Known from east to west coast and Tasmania. Some variations in colour and number of tail rings and dorsal fin rays between populations that are probably sub-specific. Primarily an estuary species that is highly localised, occupying algae reefs and seagrass beds, depending on the area. In New South Wales mainly intertidal seagrass beds and Victoria on algae reefs in 6 or more m. depth. In Tasmania to 25 m depth. Length to 20 cm.

A

A

V. philippi. Portsea Pier, Victoria, Australia. Female, length 12 cm.

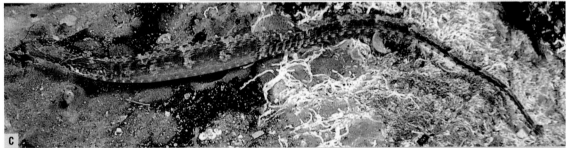

B

V. philippi. Wallaga Lake, New South Wales, Australia. Length about 14 cm.

C

V. philippi. Portsea Pier, Victoria, Australia. Female, length 18 cm.

Australian Long-snout Pipefish
Vanacampus poecilolaemus

Syngnathus poecilolaemus Peters, 1869.
Adelaide, South Australia.

Several separate populations along Australia's south coast. Mainly known from the protected waters of the S.A gulfs and Kangaroo Island, where on shallow algae reefs and in seagrass beds. W.A. population with numerous small dark spots may represent a separate species. Recognised by the long snout. Length to 28 cm.

A

V. poecilolaemus. Pelican Lagoon, Kangaroo Island, South Australia. Length 18 cm.

B

V. poecilolaemus. Kangaroo Island, South Australia. Gravid female. Length 22 cm.

Syngnathus vercoi Waite & Hale, 1921.
Spencer Gulf, South Australia.

Only known from Pelican Lagoon, Kangaroo Island and old record from Spencer Gulf. Presently only know to occur commonly in Pelican Lagoon and is vulnerable with such a small localised and special habitat. Usually amongst shallow seagrasses in tidal channels near intertidal zone. Length to 11 cm.

A

V. vercoi. Pelican Lagoon, Kangaroo Island, South Australia. Female, length 10 cm.

B

C

V. vercoi. Pelican Lagoon, Kangaroo Island, South Australia. Male, length 11 cm.

GENUS *NOTIOCAMPUS* Dawson, 1979

Masculine. Type species *Nannocampus ruber* Ramsay & Ogilby, 1886. Monotypic, sub-temperate Australian genus. One of the least known species with few specimens known. Male not known at all. Any additional information on this species is welcomed by the author.

Red Pipefish *Notiocampus ruber*

Nannocampus ruber Ramsay & Ogilby, 1886.
Sydney Harbour, Australia.

Only known from a few specimens, but ranging from Sydney to southern WA and Tasmania. Reported from 5–20 m depth. Specimen in photograph was found in 20 m depth amongst filamentous red algae and was moved to enable photograph to be taken. Moves snake-like and quickly, easily loosing sight of it. No others were found during intensive searches on consecutive dives. Length to 17 cm.

A

N. ruber. Bicheno, Tasmania, Australia. Female, length 12 cm.

B

N. ruber. Bicheno, Tasmania, Australia. Female, length 12 cm.

Masculine. Type species *Lissocampus caudalis* Waite & Hale, 1921. Comprises 5 species, 4 that are confined to the Australian and New Zealand region, and one is only known from the northern Red Sea. Small secretive species that live on algae reefs. Some are locally common, including in popular diving areas, but rarely observed, except by those searching for them. Some species have been kept in captivity and do well, readily breed, but will suffer if having to compete for food with less shy species. They produce relatively large young that could easily be raised, but no attempts have been made until now. As adults they live for about 3–4 years.

Australian Smooth Pipefish
Lissocampus caudalis

Lissocampus caudalis Waite & Hale, 1921.
South Australia.

Widespread Australia's south coast. Algae reefs and on substrate in seagrass beds, mimicking exposed roots. Semi-exposed coastal bays in mixed rubble and low algae covered reefs. Adults usually in pairs. An interesting species to keep in captivity, quickly adapting to condition and readily breeds in spacious surroundings with some natural rocks. Length to 10 cm.

A · *L. caudalis.* Portsea Pier, Victoria, Australia. Female, length 10 cm.

D · *L. caudalis.* Portsea, Victoria. Female, length 10 cm.

B · *L. caudalis.* Portsea Pier, Victoria, Australia. Brooding male, length 10 cm.

C · *L. caudalis.* Flinders, Vic. Mimicking seagrass roots.

E · *L. caudalis.* Portsea Pier, Victoria, Australia. Brooding male, length 10 cm.

F · *L. caudalis.* Portsea, Victoria. Courting pair.

A

L. runa. Bicheno, Tasmania. 8 cm.

Javelin Pipefish *Lissocampus runa*
Festucalex (Campichthys) runa Whitley, 1931.
Sydney, Australia.

Widespread Australia's south coast. Reported from reefs and tidepools. Specimens in photographs were found in about 6–10 m depth amongst algae-rubble reef, but few were found with the many hours spent. This is not an indication of the species being rare, but its common habitat not yet known. Any information on this species would be appreciated by the author. Length to about 10 cm.

C

L. runa. Kangaroo I, SA. Length 8 cm.

B

L. runa. Victor Harbour, South Australia. 6 cm.

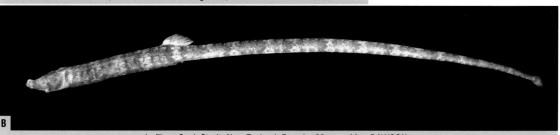

A

L. filum. Cook Strait, New Zealand. Brooding male, 95.5 mm. After DAWSON.

New Zealand's Short-snout Pipefish
Lissocampus filum
Ichthyocampus filum Günther, 1870.
Bay of Islands.

New Zealand and Chatman Islands. Lives on algae rock substrate in depths of from just below low tide level to about 10 m. Length to 11 cm.

B

L. filum. Cook Strait, New Zealand. Female, 92 mm. After DAWSON.

Prophet's Pipefish
Lissocampus fatiloquus
Campichthys fatiloquus Whitley, 1943.
Shark Bay, Western Australia.

Only known from the west coast of Australia. Was thought to be a variation of *L. caudalis* in Victoria at some stage, but it differs considerably in colour. Specimens were found on remote rocks with sargassum weeds on open substrate in 3–10 m depth. Length to about 9 cm.

A

B

C

L. fatiloquus. **A-B** Male. **C** Female. Shark Bay, Western Australia.

Suez Pipefish *Lissocampus bannwarthi*
Ichthyocampus bannwarthi Duncker, 1915.
Suez, Red Sea.

Only known from northern Red Sea and Gulf of Aqaba. Collections comprise specimens taken in few metres depth from rubble bottom. Length to about 15 cm.

A

L. bannwarthi. Gulf of Aqaba, Red Sea. Female, length 10 cm. After DAWSON.

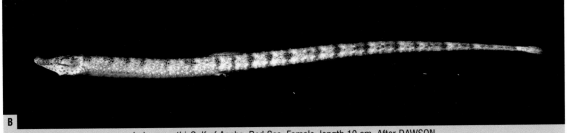

B

L. bannwarthi. Gulf of Aqaba, Red Sea. Female, length 10 cm. After DAWSON.

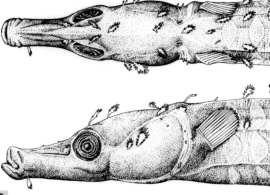

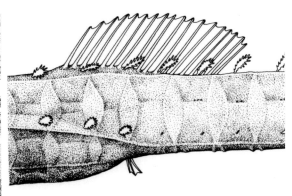

C

L. bannwarthi. Various aspects of head and body, drawn from a Red Sea specimen. Length 13 cm. After DAWSON.

Masculine. Type species *Syngnathus blainvilleanus* Eydoux & Gervais, 1837. Sub-temperate, southern hemisphere genus with 3 species, restricted to South American and New Zealand waters. Males have the brood pouch under the tail. Large females have compressed and deep trunks. Habitats are primarily estuarine and coastal waters, but reported to deep water offshore. Juveniles to about 65 mm are known from surface plankton trawls.

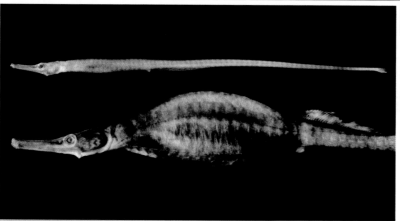

American Deep-bodied Pipefish
Leptonotus blainvilleanus

Syngnathus blainvilleanus
Eydoux & Gervais, 1837. Oceania.

Atlantic and Pacific coasts of South America. From Argentina to Chili. Most specimens are known from shallow estuaries to 25 m depth, but one specimen reported from a 250 m trawl. Length to 25 cm.

L. blainvilleanus. Top: Argentina. Male, 16 cm. Below: Chile. Female, 22 cm. After DAWSON.

High-body Pipefish
Leptonotus elevatus

Doryichthys elevatus Hutton, 1872. Wellington Harbour, New Zealand.

Widespread New Zealand. Rubble and weedy substrates. Reported from shallow bays near piers and wharfs and deep offshore between 60 and 120 m. Length 16 cm.

L. elevatus. New Zealand. Top: Brooding male, 11 cm. Below: Female, 12 cm. After DAWSON.

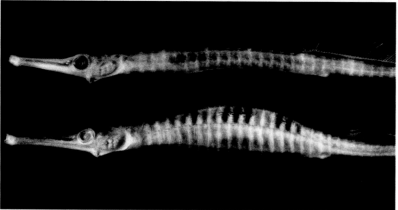

New Zealand Long-snout Pipefish
Leptonotus norae

Syngnathus norae Waite, 1910.
Steward I., to Pegasus Bay, New Zealand.

Mainly known from southern New Zealand but as far north as Hawkes Bay, North Island. Mainly known from offshore in depths from 40 to 200 m. Length to 22 cm.

L. norae. New Zealand. Top: Sub-adult, 13 cm. Below: Female, 18 cm. After DAWSON.

Masculine. Type species *Leptonotus costatus* Waite & Hale, 1921. Sub-temperate Australian monotypic genus restricted to southern waters. Occurs mostly in estuaries, associating with *Zostera* seagrass beds. Males have a tail pouch with thick skin and folds along ventral centre. The flaps overlap and interlock, completely protecting the brood.

Because it is so habitat specific, it is not recommended for captivity and should *not be collected*.

Deep-bodied Pipefish
Kaupus costatus

Leptonotus costatus Waite & Hale, 1921. Spencer Gulf, South Australia.

Mainly known from South Australia, but small populations in Victoria and Flinders Island in Bass Strait. Associates with *Zostera* seagrass beds and decline of habitat has made this species rare, except at American River, Kangaroo Island where it remains common due to the large healthy areas of seagrass beds. Intertidal to a few metres depth. In small aggregations. Length to 14 cm.

The presence of populations of this species indicates a healthy environment.

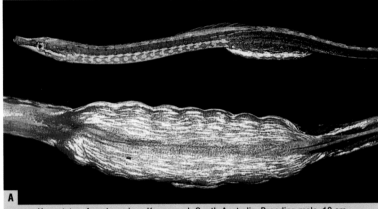

A *K. costatus*. American river, Kangaroo I, South Australia. Brooding male, 10 cm.

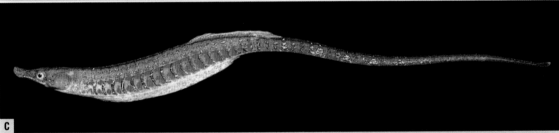

B *K. costatus*. American river, Kangaroo I, South Australia. Female, 11 cm.

C *K. costatus*. Chineman's Creek, near Port Augusta, South Australia. Female, 12 cm.

D *K. costatus*. Cape Jervis, South Australia. Female, 14 cm.

Masculine. Type species *Syngnathus tuckeri* Scott, 1942. Sub-temperate Australian genus restricted to southern waters with 4 species. They associate with densely vegetated rocky reefs or tall *Zostera* seagrasses just below the low tide mark. Males have a tail pouch with thick skin and folds along ventral centre. The flaps overlaps interlock and completely protect the brood.

Some are habitat specific, and they are not recommended for captivity. *Not be collected.*

Mollison's Pipefish
Mitotichthys mollisoni
Syngnathus mollisoni Scott, 1955.
Bivouac Bay, Tasmania.

Only known from a few specimens. The type specimen, caught accidentally in a fishing line from almost 50 m depth, and some individuals that were found at Portsea, Victoria in 7 m depth amongst brown weed on semi-open sandy substrate with sandstone rubble. Length of largest female is 22 cm.

M. mollisoni. Portsea, Victoria, Australia. Male 20 cm.

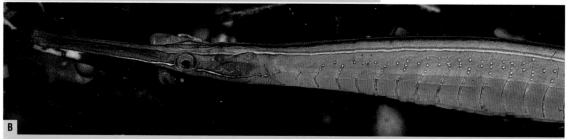

M. mollisoni. Portsea, Victoria, Australia. Female 22 cm.

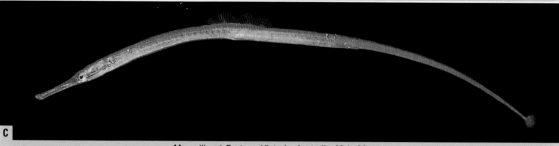

M. mollisoni. Portsea, Victoria, Australia. Male 20 cm.

M. mollisoni. Portsea, Victoria, Australia. Female 22 cm.

Half-banded Pipefish
Mitotichthys semistriatus

Leptonotus semistriatus Kaup, 1853.
Type-locality unknown.

Southern Victoria, Bass Strait region, and Tasmania. Associates with the eel-type *Zostera* seagrass beds and distribution sporadic and localised. Prefers tall seagrasses in very protected areas, usually in groups just below inter-tidal zone to about 3 m depth. Length to 25 cm.

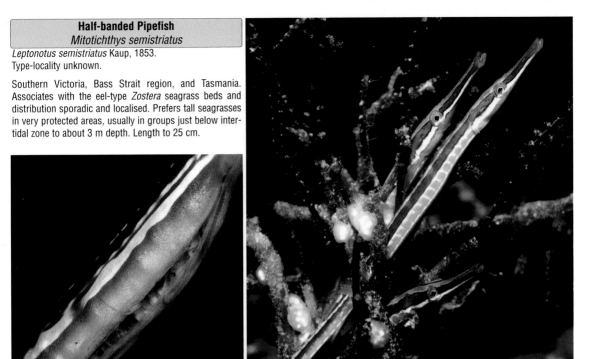

M. semistriatus. Sorrento, Victoria, Australia. **A** pouch of male opening to release young. **B** Males 22 cm.

M. semistriatus. Sorrento, Victoria, Australia. Female 24 cm.

Additional species of *Mitotichthys*

Tucker's Pipefish *Mitotichthys tuckeri.* As *Syngnathus tuckeri* Scott, 1942. Bridport Tasmania.
Only known from a few specimens, one from Twofold Bay, New South Wales, and others from Tasmania. Reported from 9–20 m depth, living in Kelp and floating Sargassum weeds. Length to about 17 cm.

Western Crested Pipefish *Mitotichthys meraculus.* As *Histiogamphelus meraculus* Whitley, 1948. Perth, Western Australia.
A doubtful species, only known from a few females from Flinders Bay and Perth, Western Australia, largest 23 cm long. Sub-adult females that were provisionally placed in the presently used genus by Dawson on the basis of the nose-ridge being absent. This now seems incorrect and it is highly possible that they belong in the original genus and, in addition, maybe a range extension for *H. briggsi.* Fresh material is needed to determine this further.

Masculine. Type species *Histiogamphelus briggsi* McCulloch, 1914. Sub-temperate Australian genus restricted to southern waters with 2 species, with eastern and western distributions. Open sand substrates with loose weeds. May migrate as adults.

A

B

H. briggsi. Portsea, Victoria, Australia. Female, 25 cm.

Brigg's Crested Pipefish
Histiogamphelus briggsi

Histiogamphelus briggsi McCulloch, 1914. Tasmania, Australia.

Mainly the Bass Strait region, ranging to Seal Rocks, New South Wales and eastern South Australia. Length to 14 cm. Occurs along reefs and off beaches with loose weeds on sand and appears to be on the move most of the time. Often seasonal in certain areas. This seems to suggest some migration behaviour. Usually in depth over 10 m off beaches, but shallower in estuaries where in sand channels. Length to 25 cm.

C

H. briggsi. Bondi Beach, Sydney, Australia. Females and male in front, about 22 cm.

H. briggsi. Seal Rocks, NSW. Juv, 5 cm.

D

H. briggsi. Portsea, Victoria, Australia. Brooding male, 25 cm.

A

H. cristatus. Rapid Bay, South Australia. Female, about 16 cm.

Rhino Pipefish
Histiogamphelus cristatus

Leptoichthys cristatus Macleay, 1882, Western Australia.

South Australian and southern Western Australia. Occurs in sparse seagrasses that border onto open sand and rubble substrates. Juveniles usually amongst loose weed that accumulates on sand by currents, providing a form of transport. Juveniles are readily identified by the 'Rhino' hump on the snout. Length to 25 cm.

B

H. cristatus. Victor Harbour, South Australia. Juvenile, about 10 cm.

C

H. cristatus. Vic. H., S. A. Juv, about 4 cm.

Masculine. Type species *Ichthyocampus cristatus* McCulloch & Waite, 1918. Sub-temperate Australian genus restricted to southern waters, and represented by a single species. Males have a fully enclosing pouch underneath the anterior section of the tail.

Ring-back Pipefish *Stipecampus cristatus*

Ichthyocampus cristatus McCulloch & Waite, 1918.
South Australia.

Bass Strait region to South Australia. Moves into breeding areas in September. Usually found in sparse seagrasses near tidal channels in large estuaries. Length to 25 cm.

Adapts well to a large aquarium and young are easily raised.

A

S. cristatus. Portsea, Victoria, Australia. Length 15 cm.

B

S. cristatus. Portsea, Victoria, Australia. Brooding male, 22 cm.

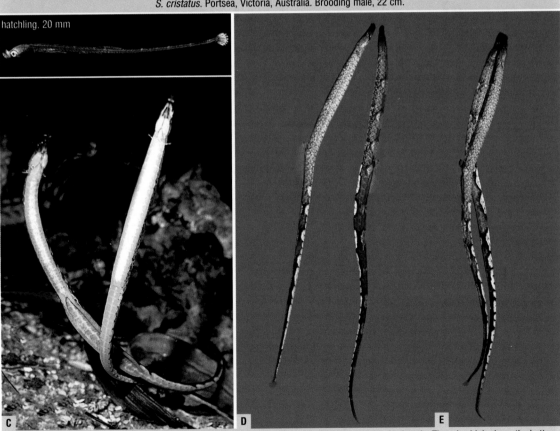

hatchling, 20 mm

C　　**D**　　**E**

S. cristatus. Portsea, Victoria. Courtship and mating. Male displaying open pouch after courtship by female. They rise high above the bottom so the female can deposit the brood into the male's pouch with the waiting sperm. Eggs are laid in a sheet that unrolls as it is pushed onto the open pouch. The female often pushes herself against the male whilst twisting her body, assisting in putting the eggs into place. Each egg needs to become embedded into the soft wall of the pouch that contains the sperm. Incubation of the about 100 eggs takes just over 4 weeks.

GENUS *LEPTOICHTHYS* Kaup, 1853

Masculine. Type species *Leptoichthys fistularius* Kaup, 1853. Sub-temperate Australian genus, monotypic. Associates with seagrasses. Large species, not recommended for captivity.

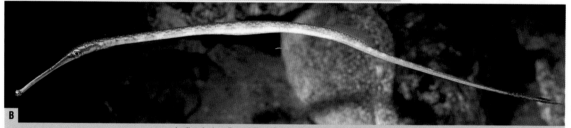

A

L. fistularius. Rapid Bay, South Australia. Length 45 cm.

Brushtail Pipefish
Leptoichthys fistularius

Leptoichthys fistularius Kaup, 1853.
King Georges Sound, Western Australia.

Various locations on the southern Australian coast and Tasmania. Inhabits large open seagrass fields near open sea. Mainly found in broad-leaved *Zostera* seagrasses where it has excellent camouflage. Juveniles swim in small groups, sometimes seen when moving through open patches. The largest known pipefish that can reach 65 cm, plus the caudal fin that is long and often elongated.

B

L. fistularius. Rapid Bay, South Australia. Length 45 cm.

C

L. fistularius. Esperance, south Western Australia. Length 50 cm.

GENUS *KIMBLAEUS* Dawson, 1980

Masculine. Type species *Kimblaeus bassensis* Dawson, 1980. Sub-temperate Australian genus, monotypic. Open sand substrates with mixed invertebrates at moderate depths.

K. bassensis. Port Lincoln, South Australia. Brooding male, 20 cm. Neville COLEMAN.

Trawl Pipefish
Kimblaeus bassensis

Kimblaeus bassensis Dawson, 1980.
Bass Strait, Victoria, Australia.

Only known from a few specimens from the Bass Strait region, Bruny Island, and the included photograph from near Port Lincoln, South Australia. Open coast species on rubble substrates, well camouflaged, and is rarely seen by divers. Known depth range 10 to 75 m. Length to 20 cm.

Masculine. Type species *Histiogamphelus rostratus* Waite & Hale, 1921. Sub-temperate Australian genus, two species. Adults deep offshore and mainly known from trawls. Juveniles have a long pelagic stage and may periodically drift into large estuaries with currents.

Knife-snout Pipefish
Hypselognathus rostratus

Histiogamphelus rostratus Waite & Hale, 1921.
Spencer Gulf, South Australia.

Broadly distributed from the Bass Strait region to South Australia. Juveniles to about 15 cm long are not uncommon in surface water with large jellies, when oceanic waters run into Port Phillip Bay, Victoria. Adults are rarely seen but are regular visitors on sand flats near the shore off Victor Harbour, South Australia in about 10 m depth. Length to 40 cm.

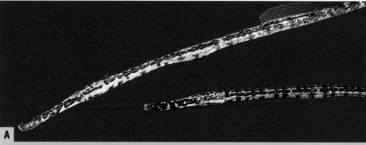

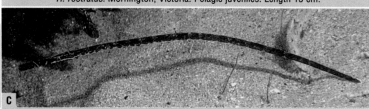

A   *H. rostratus*. Mornington, Victoria. Pelagic juveniles. Length 15 cm.

B   Mornington. Pelagic juvenile. Length 8 cm.

C   *H. rostratus*. Victor Harbour, South Australia. Length 40 cm.

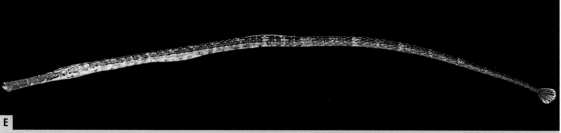

D   *H. rostratus*. Mornington, Victoria. Settled juvenile. Length about 25 cm.

E   *H. rostratus*. Mornington, Victoria. Length about 30 cm.

Additional species of *Hypselognathus*

Shaggy Pipefish *Hypselognathus horridus* Dawson & Glover, 1982. Great Australian Bight.
Only known from trawled specimens in the Great Australian Bight, taken at depths of 40–55 m. Reported at being greyish with numerous tiny brown or black spots and rings with dusky margins. Length to 28 cm.

Masculine. Type species *Choeroichthys valencienni* Kaup, 1856. Tropical Indo-West Pacific genus with at least 6 species. Males have a simple brood pouch under the trunk. Small, secretive fishes, living in the cover of shallow reefs and vegetation, and are rarely observed by divers unless looked for. Little is known about aquarium behaviour, but they are probably easily kept and would make interesting subjects like some of the small and secretive temperate species do.

A

C. brachysoma. Kerama Japan. Length about 45 mm. Atsushi ONO.

B

C. brachysoma. Mabul, Malaysia. Length 50 mm.

Pacific Short-bodied Pipefish
Choeroichthys brachysoma

Syngnathus brachysoma Bleeker, 1855.
Batu Archipelago, Indonesia.

West Pacific. Uncertain distribution with many erroneous records throughout the Indo-West Pacific. Snout length highly variable, causing some confusion with similar species from Indian Ocean, and possibly localised forms, as one shown in the photograph from Japan, which maybe additional species. **B** represents true species. Dark brown with parallel of small black spots in series along trunk, and some white speckles. Shallow reef and seagrass habitats to about 20 m depth. Length to 65 mm.

A

C. valencienni. Inhaca, southern Mozambique. Length 42 mm. Phil HEEMSTRA.

B

C. valencienni. Rowley Shoals, W. Australia. Length 45 mm. Roger C. STEENE.

Indian Short-bodied Pipefish
Choeroichthys valencienni

Choeroichthys valencienni Kaup, 1856.
Réunion.

West Indian Ocean and Red Sea. Previously included with *C. brachysoma* but snout is generally longer. Dark brown with irregular fine dark streaks on head and trunk, and often with two series of small pale ocelli along trunk, as in *C. sculptus*. Tail with thin pale bars, one following each segment. Length to about 50 mm

C. latispinosus. Dampier Archipelago, W. Australia. Length 40 mm. Barry Hutchins.

Muiron Island Pipefish
Choeroichthys latispinosus

Choeroichthys latispinosus Dawson, 1978.
South Muiron Island, Western Australia.

Only known from a 29 mm female, that was collected at 8 m on a reef slope, and included photograph. Similar to *C. brachysoma*, but deeper snout with lateral spines at tip. Tail rather thick and with pale bands. Rubble reef slopes to about 10 m depth. Length to 50 mm.

Sculptured Pipefish
Choeroichthys sculptus

Doryichthys sculptus Günther, 1870. Fiji.

Apparently widespread Indo-West Pacific, probably species-complex. Short-snouted. Dark, blackish brown with small black spots and small white ocelli in parallel series along trunk, and some white speckles over the back. Shallow reef and seagrass habitats to few metres depth. Length to 80 mm.

C. sculptus. Christmas I, Indian Ocean. Length 65 mm. Roger C. STEENE.

Pig-snouted Pipefish
Choeroichthys suillus

Choeroichthys suillus Whitley, 1951. Queensland, Australia.

Tropical Australia to southern New Guinea. Secretive on reef flats under rubble pieces to about 15 m depth. Similar to *C. brachysoma* but much more slender. Colour variable from brown to near black or yellow, with several pale blotches over back and scattered on sides. Length to 80 mm.

A *C. suillus*. Moreton Bay, Queensland. Female, 65 mm.

B *C. suillus*. Dampier Archipelago, Western Australia. Lower one, brooding male. Barry Hutchins

Barred Short-bodied Pipefish
Choeroichthys cinctus

Choeroichthys cinctus Dawson, 1976. Moluccas Is. Indonesia

Widespread Indonesia, to Queensland, Australia and to Samoa. Pale white to yellowish with distinct black barring on the trunk, followed by spots on the tail. Sheltered reef habitats, usually in crevices with sponges and in depths over 10 m. Length to 80 mm.

A *C. cinctus*. Maumere, Flores, Indonesia. Length 40 mm.

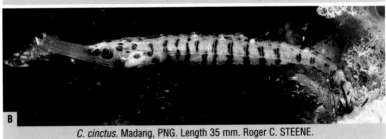

B *C. cinctus*. Madang, PNG. Length 35 mm. Roger C. STEENE.

List of other described species of *Choeroichthys*

Regarded as a synonym of *C. sculptus*:
MacGregor's Pipefish *Choeroichthys macgregori*. As *Doryrhamphus macgregori* Jordan & Richardson, 1908. Philippines.
Stumpy Pipefish *Choeroichthys ocellatus*. As *Microphis ocellatus* Snyder, 1909. Tanegashima I, Japan.

Regarded as a synonym of *C. brachysoma*:
Dotted-line Pipefish *Choeroichthys serialis*. As *Doryichthys serialis* Günther, 1884. Queensland, Australia. (=*suillus?*)

Short-fin Pipefish *Choeroichthys smithi*. Dawson, 1978. Inhaca, Mozambique. Similar to *C. latispinosus*.

GENUS *BHANOTIA* Hora, 1925

Feminine. Type species *Syngnathus corrugatus* Weber, 1913 (= *S. fasciolatus* Duméril, 1870). Tropical Indo-West Pacific genus with 3 small species. Males have a brood pouch under the tail and young may have a long pelagic stage, reaching near adult size.

Small, secretive fishes, living in the cover of shallow algae reef habitats, in tidal pools and sheltered reef flats.

Corrugated Pipefish *Bhanotia fasciolata*

Syngnathus fasciolatus Duméril, 1870.
Java Indonesia.

Known from Andaman Sea to New Hebrides. Head raised with dorsal ridge from above posterior edge of eye. Indistinct dark bars below head. Lives openly on muddy or silty substrates in 3 to 25 m depth. Length to 9 cm.

Bhanotia sewelli Hora, 1925, from Andaman Sea, *Syngnathus uncinatus* and *S. corrugatus* Weber, 1913 from Indonesia are regarded as synonyms.

A
B. fasciolata. Phuket, SE coast, Thailand. Length 80 mm. Ukkrit SATAPOOMIN.

B
B. fasciolata. Phuket, SE coast, Thailand. Length 80 mm. Ukkrit SATAPOOMIN.

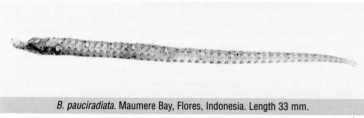

Few-rayed Pipefish *Bhanotia pauciradiata*

Bhanotia pauciradiata Allen & Kuiter, 1995.
Maumere, Flores, Indonesia.

Known from single specimen, collected from a reef slope with algae covered coralline rock-rubble. With small size and excellent camouflage it is easily overlooked. Length only 33 mm and probably a fully grown female.

B. pauciradiata. Maumere Bay, Flores, Indonesia. Length 33 mm.

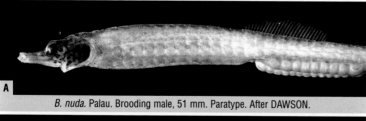

Naked Pipefish *Bhanotia nuda*

Bhanotia nuda Dawson, 1978. Palau.

Only known from Palau and south-eastern Papua New Guinea. To about 55 mm.

A
B. nuda. Palau. Brooding male, 51 mm. Paratype. After DAWSON.

B
B. nuda. Milne Bay, PNG. Male, 51 mm. After DAWSON.

C
B. nuda. Head detail of type. After DAWSON.

Masculine. Type species *Micrognathus brachyrhinus* Herald, 1953, as subgenus of *Micrognathus* Duncker. Circum-tropical genus with 3 or 4 similar small species divided over various areas. Males have a brood pouch under the tail and young may have a long pelagic stage, reaching near adult size. Small, secretive fishes, living in the cover of shallow algae reef habitats, in tidal pools and sheltered reef flats to deep water, depending on species.

Myer's Pipefish
Minyichthys myersi

Micrognathus (Minyichthys) myersi
Herald & Randall, 1972. Guam.

Widespread West Pacific and specimens from western Indian Ocean are thought to represent this species also. Rubble reef slopes and flats at various depths from about 6 m. Length to 60 mm.

M. myersi. Maumere Bay, Flores, Indonesia. Length 40 mm.

West Atlantic Pipefish
Minyichthys inusitatus

Minyichthys inusitatus. Dawson, 1983. Western Atlantic.

Only known from the Western Atlantic.

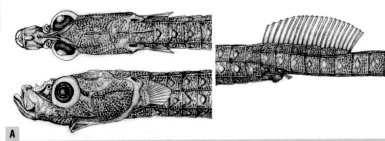

A

M. inusitatus. Caribbean. Length 29 mm SL. After DAWSON.

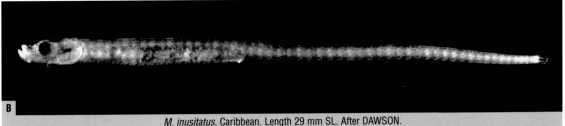

B

M. inusitatus. Caribbean. Length 29 mm SL. After DAWSON.

East Atlantic Pipefish
Minyichthys sentus

Minyichthys sentus. Dawson, 1982. Eastern Atlantic.

Only known from the Eastern Atlantic.

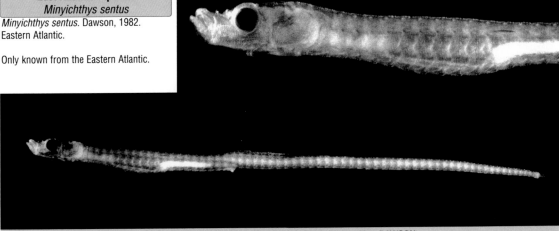

M. sentus. East Atlantic. Length 48 mm SL, holotype. After DAWSON.

Masculine. Type species *Ichthyocampus tryoni* Ogilby, 1890, as subgenus of *Festucalex* Whitley. Indo-Pacific genus with 4 similar small described species, and at least 3 more that are undescribed, based on juveniles and planktonic specimens. 3 of the described species are Australian endemics. Little known species, usually dredged from rubble substrate. Brood pouch is under the tail.

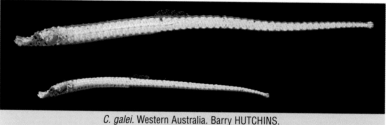

C. galei. Western Australia. Barry HUTCHINS.

Gale's Pipefish *Campichthys galei*

Ichthyocampus galei Duncker, 1909. Freycinet Estuary, Shark Bay, Western Australia.

Restricted to south-western Australian Waters, ranging to Shark Bay in the north and Great Australian Bight in the south. Found on shallow rubble substrate to 18 m depth. Length to 60 mm.

C. tryoni. Moreton Bay, Qld. Female above, male below, about 60 mm. After DAWSON.

A

B

C. tryoni. Moreton Bay, Qld. Depth 10 m. Female, length 60 mm.

Tryon's Pipefish *Campichthys tryoni*

Ichthyocampus tryoni Ogilby, 1890. Moreton Bay, Queensland, Australia.

Only known from Queensland, Australia. On shallow reef flats and reef margins bordering on to sand channels. Usually secretive under large rubble pieces. Dredged on rubble substrates in estuaries and inner reefs. Length to 75 mm.

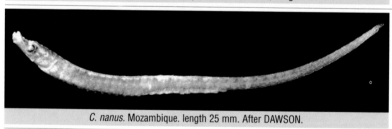

C. nanus. Mozambique. length 25 mm. After DAWSON.

Pixy Pipefish *Campichthys nanus*

Campichthys nanus Dawson, 1977. Pinda, Mozambique.

Only known from 2 specimens near type-locality. One of the smallest tail-pouch species, maturing at 25 mm., the size of both specimens.

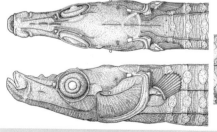
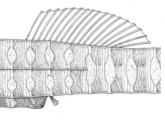

C. tricarinatus. Northern Australia. After DAWSON.

Three-keel Pipefish *Campichthys tricarinatus*

Campichthys tricarinatus. Dawson, 1977. Baleine Bank, Western Australia.

Northern Australian waters. Only known from 4 specimens, no adult males known, 22–44 mm SL. Rubble and remote coral heads.

Masculine. Type species *Syngnathus brevirostris* Rüppell, 1838. Indo-Pacific and Western Atlantic genus with at least 6 species. One is found off Brazil. Males have a brood pouch under the tail.

Small, secretive fishes, living in the cover of large rubble reef habitats.

Thorn-tailed Pipefish
Micrognathus brevirostris

Syngnathus brevirostris
Rüppell, 1838. Red Sea.

Only known from the Red Sea, Gulfs of Suez and Aqaba. In algal and coral habitats to about 10 m depth. Length to 70 mm

M. brevirostris. Egypt, Red Sea. Male 45 mm, female 50 mm. After DAWSON.

Anderson's Pipefish
Micrognathus andersonii

Syngnathus andersonii Bleeker, 1858. Kokos Is, Indonesia.

Reported as widespread Indo-West Pacific, but probably restricted to Indonesia, and part of a species complex with several geographical variations. Inhabits tidepools and usually amongst sparse seagrasses and rubble to about 10 m depth. Length to 85 mm.

A

M. andersonii. Lizard I, Queensland, Australia. Length 60 mm.

B

M. andersonii. Mabul, Borneo, Malaysia. Brooding male, 65 mm.

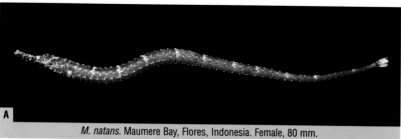

Offshore Pipefish
Micrognathus natans

Micrognathus natans Dawson, 1982. Fiji.

West Pacific, Philippines to Queensland, Australia and Fiji. Similar to *M. andersonii,* but more slender, longer snout and has fewer trunk-rings, normally 14 versus 15–17. Most specimens are known from plankton collections. Specimens from Flores were found on muddy, silty substrates in 15–25 m depth and may prefer deep sheltered habitats. Length to 60 mm.

M. natans. Maumere Bay, Flores, Indonesia. Female, 80 mm.

M. natans. Maumere Bay, Flores, Indonesia. Brooding male, 85 mm.

Pygmy Pipefish
Micrognathus pygmaeus

Micrognathus pygmaeus Fritzsche, 1981. Tahiti.

Widespread West Pacific to central Pacific. Closely related *M. brevirostris* in the Red Sea. Sheltered inner reefs in large lagoons. In narrow crevices and small caves to about 10 m depth. Length to 60 mm.

M. pygmaeus. Guam. Female, about 45 mm. Rob MYERS.

M. pygmaeus. Kerama, Japan. Female, about 45 mm. Atsushi ONO.

List of other described species of *Micrognathus*

Brazilian Pipefish *Micrognathus erugatus* Herald & Dawson, 1974. Bahia, Brazil.
Endemic to Brazil.
Tidepool Pipefish *Syngnathus micronotopterus* Fowler, 1938. Philippines.
West Pacific, southern Japan to north-western Australia. Males have several large black spots on upper trunk, following white saddle markings. Inhabits tidepools and usually amongst sparse seagrasses and algae-rubble to about 6 m depth. Length to 70 mm.
Tanaka's Pipefish *Corythroichthys tanakai.* Jordan & Starcks, 1906. Tanegashima, Japan. Thought to a synonym of *Micrognathus andersonii.*

Masculine. Type species *Corythoichthys albirostris* Kaup, 1856. Circum-tropical genus with about 14 species, but there is a wide variety in gross morphology and some species are known from a few specimens. Much more information is needed to solve the taxonomic problems.

White-nose Pipefish
Cosmocampus albirostris
Corythoichthys albirostris
Kaup, 1856. Mexico, Western Atlantic.

West Atlantic, Florida, Bahamas and Caribbean seas. Easily recognised by its white snout. Rubble and sparse algae habitat to about 40 m depth. Length to 20 cm.

C. albirostris. St. Vincent, Caribbean. Female, 12 cm. Roger C. STEENE.

C. albirostris. St. Vincent, Caribbean. Female, 12 cm. Roger C. STEENE.

Snub-nose Pipefish
Cosmocampus arctus
Siphostoma arctum
Jenkins & Evermann, 1889.
Guaymas, México.

East Pacific, central California to Mazatlán, México. Mixed reef and seagrass habitat to 10 m depth. Length to 13 cm.

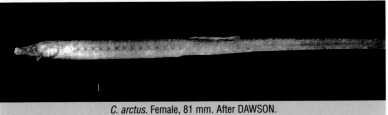

C. arctus. Female, 81 mm. After DAWSON.

Herald's Pipefish
Cosmocampus heraldi
Bryx (Simocampus) heraldi Fritzsche, 1980. San Felix I, Chile.

Only known from few specimens taken from rocky reefs and sand bottom in 6–23 m at type locality. Length to 12 cm.

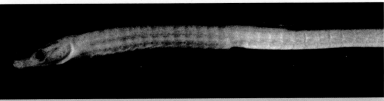

C. heraldi. Chile. Male, 70.5 mm, paratype. After DAWSON.

Lord Howe's Pipefish
Cosmocampus howensis
Parasyngnathus howensis Whitley, 1948.
Lord Howe Island.

Appears to be widespread in the south Pacific. *Syngnathus caldwelli* Herald & Randall, 1972, from Easter Island is regarded a synonym. Length to about 12 cm.

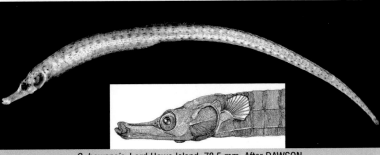

C. howensis. Lord Howe Island, 78.5 mm. After DAWSON.

A

H. elucens. Honduras, West Atlantic. Length about 12 cm. Paul HUMANN.

B

C. elucens. West Atlantic. Female, 14 cm. Paul HUMANN.

C

C. elucens. Cayman Islands, West Atlantic. Female, 15 cm. Paul HUMANN.

Short-fin Pipefish
Cosmocampus elucens

Syngnathus elucens Poey, 1868.
Havana, Cuba, Western Atlantic.

West Atlantic, Florida, Bahamas and Caribbean seas. Slender, long-snouted species with evenly spaced pale cross bars over the back. Well camouflaged when in algaes. Various reef and seagrass habitats, but usually with mixed algae or weeds. Reported from shallow depths but venture to deep water. Length to 15 cm.

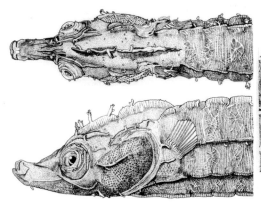

A

H. brachycephalus. Male, 86 mm. After DAWSON.

American Crested Pipefish
Cosmocampus brachycephalus

Syngnathus brachycephalus Poey, 1868.
Tortugas, Florida.

West Atlantic, Bahamas, Antilles, East Florida coast and western Caribbean. Shallow sub-tidal coastal grass flats. Length to 10 cm.

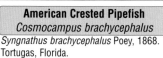

B

H. brachycephalus. Female, 69 mm. Various aspects of the head and trunk to tail part. After DAWSON.

Ball's Pipefish
Cosmocampus balli

Corythoichthys balli Fowler, 1925. Hawaii.

Hawaiian endemic. Shallow protected reef habitats. Length to 60 mm.

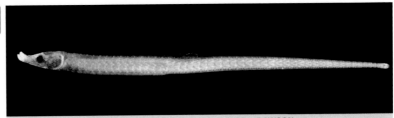

H. balli. Hawaii. Female, 68 mm. After DAWSON.

H. balli. Hawaii. Female, 68 mm. Various aspects of the head. After DAWSON.

American Dwarf Pipefish
Cosmocampus hildebrandi

Syngnathus hildebrandi Herald, 1965. Tortugas, Florida.

West Atlantic, North Carolina, SE Florida and off West Florida. Sand with coral and rock substrates between 5 and 75 m depth. Length to 85 mm.

H. hildebrandi. Florida. Male, 50.5 mm. After DAWSON.

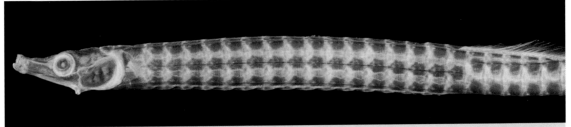

H. hildebrandi. Florida. Female, 86 mm, holotype. After DAWSON.

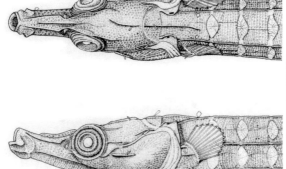

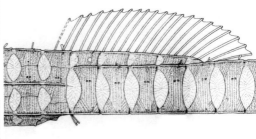

H. hildebrandi. Florida. Female, 86 mm, holotype. After DAWSON.

Scarlet Pipefish
Cosmocampus coccineus

Syngnathus coccineus Herald, 1940. Galapagos I.

East Pacific, Galapagos I. Similar *C. arctus* on mainland. Rock and coral reefs to about 20 m depth. Length to 13 cm.

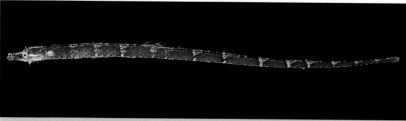

C. coccineus. Playa Venao, Panama. G.R. ALLEN.

Investigator Pipefish
Cosmocampus investigatoris

Syngnathus (?) investigatoris. Hora, 1925. Mergui Harbour, Lower Burma.

Known from Andaman Sea to Arabian Gulf. Brown with dark band on each ring. Looks very much like some of the whip-corals in the same areas where it's found. Rocky substrates in current prone areas, to 15 m depth. Length to 10 cm.

C. investigatoris. Andaman Sea. 10 cm. Mark STRICKLAND.

Maxweber's Pipefish
Cosmocampus maxweberi

Syngnathus maxweberi Whitley, 1933.
Replacement name for *S. punctatus* Weber, 1913. Sumbawa, Indonesia.

West Pacific. Secretive, inner reefs at moderate depths to about 35 m. Under large rubble pieces in coralline rubble. Length to 10 cm.

A

C. maxweberi. Maumere, Flores, Indonesia. Length 10 cm.

B

C. maxweberi. Maumere, Flores, Indonesia. Length 10 cm.

Back-finned Pipefish
Cosmocampus retropinnis

Cosmocampus retropinnis Dawson, 1982. North-east Atlantic, oceanic.

Eastern Atlantic. Poorly known, a small species to about 60 mm.

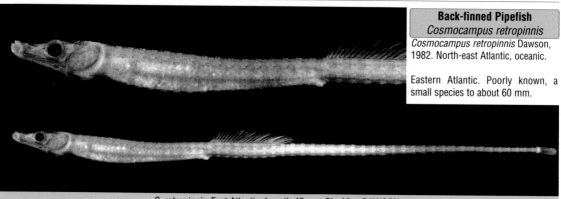

C. retropinnis. East Atlantic. Length 48 mm SL. After DAWSON.

Rough-ridge Pipefish
Cosmocampus banneri

Syngnathus banneri
Herald & Randall, 1972.
Ryukyu I, Japan.

Presently a single species recognised throughout the Indo-West Pacific, including Red Sea. Probably comprises a complex of similar species that are lumped together on the basis of similarities. Pale with some dusky bars on lower trunk, following head. Rubble reefs to 30 m depth. Length to 60 mm.

A

C. banneri. Kerama, Japan. Length 4 cm. Atsushi ONO.

B

C. banneri. Sodwana. Brooding male, length 50 mm. Phil HEEMSTRA.

D'Arros Pipefish
Cosmocampus darrosanus

Syngnathus darrosanus
Dawson & Randall, 1975.
Off D'Arros I, Amirante Islands.

West Indian Ocean with similar forms in Guam and Australia that are presently included. *Syngnathus lumbricoides* Mauge, 1981, from Madagascar regarded a synonym. Upper snout and in front of eyes bony white. Tide pools and reef flats to 3 m depth. Length to 80 mm

C. darrosanus. Tulamben, Bali, Indonesia. Takamasa TONOZUKA.

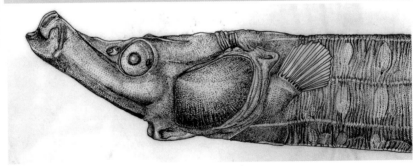

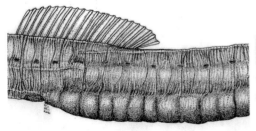

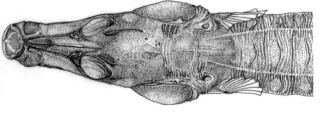

C. darrosanus. Queensland. After DAWSON.

Additional species of *Cosmocampus.*.

Bottomless Pipefish *Cosmocampus profundus,* as Corythoichthys *profundus.* Herald, 1965. East of Melbourne, Florida. Western Atlantic. West Atlantic, East Florida, Virgin Islands, and Yucatán Peninsula, México. Two specimens known from trawls over sand and coral substrates in depths of 146–265 m and a third specimen from the stomach of a fish caught in about 200 m. Length to 20 cm.

Masculine. Type species *Siphostoma crinigerum* Bean & Dresel, 1884. Similar to *Micrognathus*, but lacks anal fin. Only two species that occur in the Western Atlantic.

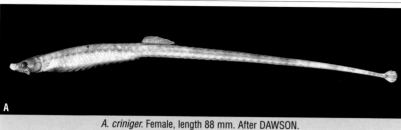

Fringed Pipefish
Anarchopterus criniger

Siphostoma crinigerum Bean & Dresel, 1884. Pensacola, Florida.

West Atlantic, Bahamas, North Carolina, SE Florida and Gulf of Mexico. Reported from seagrass beds and floating weeds. Usually in depths less than 5 m. Males brood about 50 eggs. Length to 95 mm.

A. criniger. Female, length 88 mm. After DAWSON.

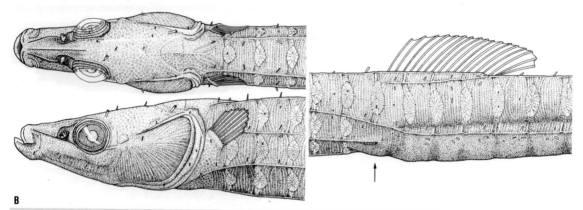

A. criniger. Male, length 88.5 mm. After DAWSON.

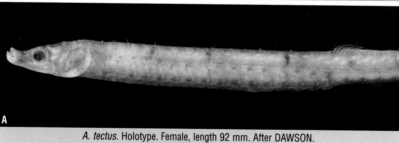

Insular Pipefish
Anarchopterus tectus

Micrognathus tectus Dawson, 1978. Alligator Reef, Florida.

West Atlantic, Bahamas, Florida Keys and several Caribbean localities south to Venezuela. Algae reefs to 25 m. Males brood over 100 eggs. Length to 12 cm.

A. tectus. Holotype. Female, length 92 mm. After DAWSON.

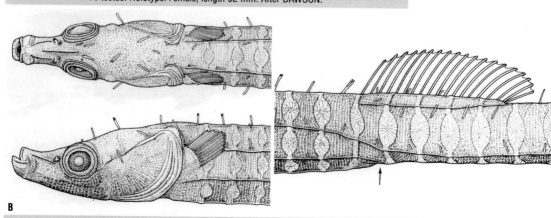

A. tectus. Holotype. Female, length 92 mm. After DAWSON.

Masculine. Type species *Siphostoma starksii* Jordan & Culver, 1895. Small genus with 3 species found in river drainages of central America and east coast of south America. Primarily freshwater streams and rivers.

Fast-river Pipefish *Pseudophallus starksii*

Siphostoma starksii Jordan & Culver, 1895.
Rio Presidio, Sinaloa, México.

Pacific drainages from Baja California, México, to Santa Rosa, Ecuador. Lives in streams with rocky bottom and vegetation with a flow rate of several kilometres per hour (slow walking pace). Male broods about 900–1600 eggs that hatch in about 10 days. Length to 18 cm.

A

P. starksii. Female, length 98 mm. After DAWSON.

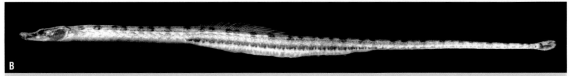

B

P. starksii. Male, length 106 mm. After DAWSON.

Freshwater Pipefish *Pseudophallus mindii*

Syngnathus mindii Meek & Hildebrand, 1923.
Near Mindi, Canal Zone, Panama.

Widespread along eastern central and south American coast from Panama to Rio Registro de Igiuape, Brazil, and several localities in the Antilles. Mainly freshwater, but can be washed out to sea by large rivers over great distances in floating weeds. Length to 16 cm.

A

P. mindii. Female, length 88 mm. After DAWSON.

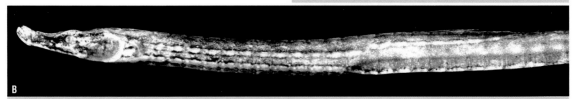

B

P. mindii. Male, length 88.5 mm. After DAWSON.

El Captain River Pipefish *Pseudophallus elcapitanensis*

Siphostoma elcapitanense Meek & Hildebrand, 1914. El Capitan, Panamá.

Pacific drainages from the vicinity of Jiménez, Costa Rica to Rio Chico, tributary of Rio Tuira, Panamá Freshwater drainages well above tidal influences. Quiet water along river banks. Usually among organic debris. Length to 20 cm.

A

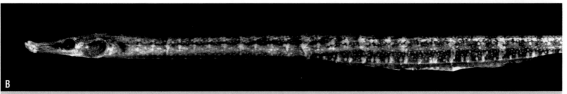

B

P. elcapitanensis. Male, length 12 cm. After DAWSON.

Pseudophallus brasiliensis Dawson, 1974. Rio Tocantins, Igarapé Inó, Faro de Panaquera, Pará, Brazil.

Atlantic drainages of Brazil. Usually among organic debris. Length to about 15 cm.

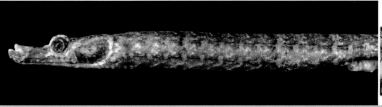

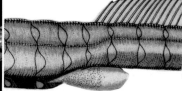

P. brasiliensis. Length ~10 cm. After DAWSON.

GENUS *NEROPHIS* Rafinesque, 1810

Masculine. Type species *Syngnathus ophiodon* Linnaeus, 1758. Small, *Syngnathus*-like genus lacking caudal fin, with 3 species in the eastern Atlantic, Mediterranean and Black Seas. Male brood eggs under trunk without pouch protection.

Atlantic Worm Pipefish *Nerophis lumbriciformis*

Syngnathus lumbriciformis Jenyns, 1835. Cornwell, England.

Atlantic European coast, British Isles, and Norway from Bergen to the Kattegat. Intertidal to about 30 m depth. Low on reef among rocks, or in holdfasts of weeds. Males brood about 120–150 eggs. Length to 17 cm.

A B

N. lumbriciformis. Roscoff, Bretagne, France. **A** female in background and male with brood in foreground. **B** female. Patrick LOUISY.

C

N. lumbriciformis. Roscoff, Bretagne, France. Male with brood. Patrick LOUISY.

Syngnathus ophiodon Linnaeus, 1758. Europe.

Widespread European coast, including most of the British Isles, from Trondheim, Norway to throughout the Mediterranean and Black Sea. Estuarine, in algae or seagrasses, usually shallow, ranging to a depth of about 15 m. Sometimes enters freshwater. Males brood about 120–150 eggs. Length to 30 cm, usually to 25 cm.

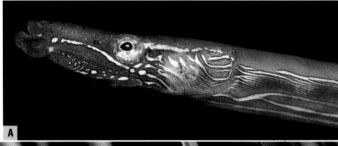

A

B

N. ophidion. Golfe du Lion, Mediterranean Sea, France. **A** female. **B** male with brood. Patrick LOUISY.

C

N. ophidion. Near Wismar, Baltic Sea. Depth ~3m. Length ~20 cm. Werner FIEDLER.

European Spotted Pipefish *Nerophis maculatus*

Nerophis maculatus Rafinesque, 1810. Palermo, Sicily.

Coastal waters of the Mediterranean and adjacent Atlantic waters. Common in Adriatic Sea. Male broods about 180 eggs. Length to 30 cm.

A

N. ophidion. Near Wismar, Baltic Sea. Depth ~3m. Length ~20 cm. Werner FIEDLER.

B

N. ophidion. Near Wismar, Baltic Sea. Depth ~3m. Length ~20 cm. Werner FIEDLER.

Masculine. Type species *Syngnathus aequoreus* Linnaeus, 1758. Small, *Syngnathus*-like genus, lacking caudal fin, with a single species in the eastern Atlantic. Male broods eggs under trunk without pouch protection.

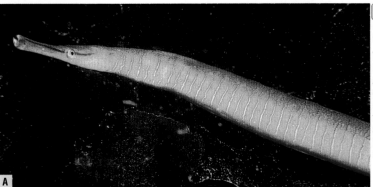

Snake Pipefish *Entelurus aequoreus*

Syngnathus aequoreus Linnaeus, 1758.
Open Sea, Europe.

Widespread European Atlantic coast, including British Isles, south Iceland, and south to Azores. Oceanic and inshore in algae or seagrasses. Adults have a large dorsal fin, but no others fins. Juveniles to about 70 mm have pectoral fins and a rudimentary caudal fin. There is a considerable difference in maximum size between sexes. Females largest, to 60 cm and males to 40 cm.

A

E. aequoreus. Isle of Man, Irish Sea. Florian GRANER.

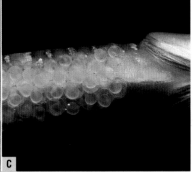

B

C

E. aequoreus. Roscoff, Bretagne, France. Male with brood. Patrick LOUISY.

GENUS *NANNOCAMPUS* Günther, 1870

Masculine. Type species *Nannocampus subosseus* Günther, 1870. Comprises 4 or 5 species variously distributed in the Indo-West Pacific. Pectoral fins present in young but only some adults. Brood pouch under tail, formed by plates and folds. Secretive in algaes and reefs.

Dampier Pipefish *Nannocampus* sp 1

Only recently discovered in Western Australia on the Dampier Archipelago. Appears to be undescribed. Closely related to *N. pictus* and similar to the next species, *N. subosseus,* and following *N. weberi* from Indonesia.

N. sp 1. Female above, male below. Dampier Archipelago, Western Australia. Barry HUTCHINS.

Bony-headed Pipefish
Nannocampus subosseus

Nannocampus subosseus Günther, 1870.
Freycinet's Harbour, Western Australia.

Only known from Western Australia, Point Demster
to Shark Bay. Mainly in rockpools and shallow reef
to 8 m depth. Length to 10 cm.

N. subosseus. Rottnest Island, Western Australia. Length 79 mm. After DAWSON.

Painted Pipefish *Nannocampus pictus*

Ichthyocampus pictus Duncker, 1915.
Gulf of Mannar, Sri Lanka.

Indian Ocean, African coast to Sri Lanka. Mixed
reef and seagrass habitat to 10 m depth. Length to
10 cm.

N. pictus. Female, length 91 mm. After DAWSON.

Elegant Pipefish *Nannocampus elegans*

Nannocampus elegans Smith, 1961.
South-eastern Africa.

South-eastern Africa, north to Mozambique. Snout
very short. Tan to brown without obvious banding
or spotting. Adults lack pectoral fins. Tide pools
and shallow reefs. Length to 12 cm

N. elegans. Inhaca, southern Mozambique. Length 46 mm. Phil HEEMSTRA.

Lindeman Pipefish
Nannocampus lindemanensis

Campichthys tryoni lindemanensis Whitley, 1948.
Lindeman I, Queensland, Australia.

Shallow reef flats in large lagoons. Algae-rubble
habitat. Length to 10 cm.

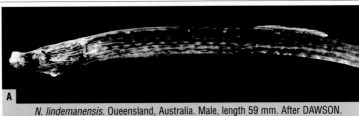

A

N. lindemanensis. Queensland, Australia. Male, length 59 mm. After DAWSON.

B

N. lindemanensis. Queensland, Australia. Male, length 59 mm. After DAWSON.

Reef-flat Pipefish *Nannocampus weberi*

Nannocampus weberi Duncker, 1915. Sumba I,
Indonesia.

Only known from Bali and Sumba, Indonesia.
Shallow reef flats between beach and barrier
reefs. Length to 10 cm.

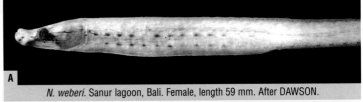

A

N. weberi. Sanur lagoon, Bali. Female, length 59 mm. After DAWSON.

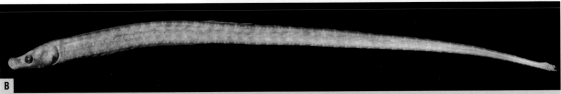

B

N. weberi. Holotype from Sumba. After DAWSON.

Masculine. Type species *Syngnathus conspicillatus* Kaup, 1856 (=. *Halicampus grayi* Kaup, 1856). Presently as a Pacific-Atlantic genus with about 14 species, but there is a wide variety in gross morphology and some species are known from a few specimens. Much more information is needed to solve the taxonomic problems.

Most species are found on mixed algae-rubble habitats and probably do well in 'living rock' aquariums, but food will be a critical part as most feed on very small crustaceans.

A

Whiskered Pipefish
Halicampus macrorhynchus

Halicampus macrorhynchus Bamber, 1915.
Suez, Red Sea.

Appears to represent a single widespread species from the Red Sea to the West Pacific. Red Sea population represents the true species. It shows slight differences with populations from elsewhere, but these maybe contributed to the environment. Colours vary with habitat, colourful on algae-rubble and dull on sand. Juveniles with round-leafed sea-grasses on sand slopes, usually settling from pelagic stage at about 8 cm long. Adults on sand or algae covered reefs to about 25 m depth. Length to 18 cm.

H. macrorhynchus. Ambon, Indonesia. Length about 15 cm. Roger C. STEENE.

B

H. macrorhynchus. Lizard Is, Queensland, Australia. Length about 8 cm.

C

H. macrorhynchus. As **E**, close-up.

D

H. macrorhynchus. Pulau Putri, Java, Indonesia. Length about 20 cm.

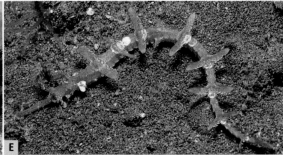

E

H. macrorhynchus. Bali, Indonesia. L. 9 cm. Takamasu TONOZUKA.

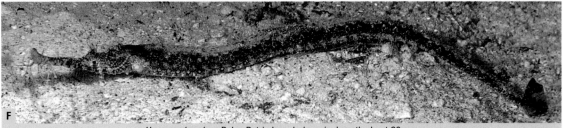

F

H. macrorhynchus. Pulau Putri, Java, Indonesia. Length about 20 cm.

H. macrorhynchus. Adult. Gulf of Aqaba, Red Sea. Thomas PAULUS.

H. macrorhynchus. Adult. Gulf of Aqaba, Red Sea. Thomas PAULUS.

H. macrorhynchus. Juvenile. Aqaba, Red Sea. Thomas PAULUS.

H. macrorhynchus. **J-K** Juvenile, **L** Brood on ventral part of the tail near the anus. Gulf of Aqaba, Red Sea. Thomas PAULUS.

A

H. brocki. Kerama, Japan. Length about 45 mm. Atsushi ONO.

Brock's Pipefish *Halicampus brocki*

Micrognathus brocki Herald, 1953. Bikini Atoll, Marshall I.

Widespread West Pacific, southern Japan to northern Australia. Slender species with branching tentacles on the head as shown in **D**. Inner reefs, coral and algae-rich habitats, usually at moderate depths. Flores specimen was photographed at 35 m depth. Length to 12 cm.

B

C

D

H. brocki. Maumere, Flores, Indonesia. Length 12 cm, male.

A

H. ensenadae. St. Vincent, Caribbean. Length 10 cm. Roger C. STEENE.

Long-haired or Harlequin Pipefish
Halicampus ensenadae

Corythroichthys ensenadae Silvester, 1915. Puerto Rico, Carribean Seas.

West Atlantic, Florida, Bahamas and Caribbean seas. Slender and highly variable species with filamentous tentacles on the head. Plain creamy to brown or strongly banded with broad pale interspaces. Shallow rocky, algae-rich habitats, to about 20 m depth. Length to 16 cm.

Remarks: The banded form is commonly called Harlequin Pipefish.

Remarks: Closely related to *H. crinitus* from Brazil (next species) and regarded as the same by some authors.

B

H. ensenadae. St. Vincent, Caribbean. Female, 12 cm. Paul HUMANN.

Brazil's Banded Pipefish
Halicampus crinitus

Syngnathus crinitus Jenyns, 1842.
Brazil, Western Atlantic.
Corythoichthys vittatus Kaup, 1856.
Brazil, Western Atlantic.

Brazil. Slender and highly variable species with or without filamentous tentacles on the head and body. Plain creamy to dark-brown or strongly banded with narrow pale interspaces. Occurs on mixed sand and algae-rocky reef in 1-7 m depth. Length to 16 cm.

Photos by Ricardo GUIMARAES

H. crinitus. Ilha Grande Bay, Brazil. **A** Female, 70 mm. **B** Female, 80 mm. **C** Male, 140 mm SL.

Red-hair Pipefish
Halicampus dunckeri

Micrognathus dunckeri Chabanaud, 1929. Ambon, Indonesia.

Indonesia and Philippines. Head with simple but long filaments and small leafy bits along the back as shown in **A**, but latter usually smaller in other individuals. Snout short with rounded dorsal ridge, and eyes large. Coastal algae-rubble slopes to 25 m depth. Length to 12 cm.

H. dunckeri. Tulamben, Bali. Depth 10-20 m. **E** Brooding male, length about 12 cm.

Red-Sea Dusky Pipefish
Halicampus sp 1

Was provisionally included with *H. dunckeri* by Dawson, 1985. Probably a Red Sea endemic, but was thought to be identical to west Indian Ocean populations. Closely related to *H. dunckeri* (see above) but also shows similarities with *H. spinirostris* with its longer snout, as well as *H. boothae* in colour and size. Lives on shallow open rubble reef flats with fine algaes. Length to about 15 cm (African population).

H. sp 1. Aqaba, Red Sea. Thomas PAULUS.

A

Undetermined species that was previously included with *H. dunckeri*, but obviously represents another taxon, and is also similar to *H. boothae*. Probably restricted to the Great Barrier Reef region. Lives on coral reefs in depths to about 20 m. Length to 14 cm.

H. sp 2. Great Barrier Reef, Australia. Female, length 11 cm. Phil WOODHEAD.

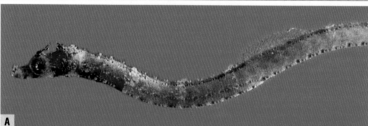

B

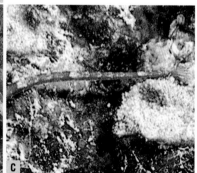

C

H. sp 2. Heron Island, Queensland, Australia. Depth 13 m. Male, length 12 cm. **C** showing caudal fin, same individual as **B**. Neville COLEMAN.

A

Spiny-snout Pipefish
Halicampus spinirostris

Micrognathus spinirostris
Dawson & Allen, 1981.
North West Cape, Western Australia.

Widespread Indo-West Pacific, but few specimens known. Pale to dark Brown with about 4 broad black bands on trunk, each separated by a thin white bar. Rubble lagoons, intertidal. Length to 12 cm.

H. spinirostris. Lipe I., off Satun province, Thailand. Length 66 mm. Ukkrit SATAPOOMIN.

B

C

H. spinirostris. Tidal lagoon, Maumere, Flores, Indonesia. Length 12 cm.

Booth's Pipefish
Halicampus boothae

Micrognathus boothae Whitley, 1964.
Lord Howe Island, Australia.

Southern West Pacific, eastern Australia and Coral Sea. Similar species in Western Indian Ocean and in Japan (next species). Very long species with evenly space pale saddles along entire length. Rocky reef with algae-rich habitats in 3–30 m depth. Length to 16 cm.

A

H. boothae. Norfolk Island. Malcolm FRANCIS.

B
H. boothae. Lord Howe Island. Neville COLEMAN.

C
H. boothae. Norfolk Island. Malcolm FRANCIS.

Long-tail Pipefish
Halicampus sp 3 (cf *boothae*)

Halicampus boothae Japanese literature.

Sub-tropical Japan to southern Philippines. Previously included with *H. boothae* that has similar colour and morphology. *H.* cf *boothae* has fewer tail rings and lower dorsal fin ray count, as well as longer snout. It lives along margins of shallow algal reefs, in sheltered bays to about 30 m depth. Length to 18 cm.

A

B

C

D
H. sp 3. Izu Peninsula, Japan. Depth 10–15 m. Length 15 cm. Hiroyuki UCHIYAMA.

A

H. grayi. Maumere, Flores, Indonesia. Depth 25 m. Length 15 cm, male.

Mud Pipefish *Halicampus grayi*

Halicampus grayi Kaup, 1856. No type-locality.

Widespread West Pacific, Andaman Sea and southern Red Sea. Short-snouted species with large eyes. Lives in muddy habitats, often covered with silt and extremely well camouflaged. Shallow inshore muddy bays to deep offshore, reported to 100 m depth. Length to 15 cm.

B

H. grayi. Maumere, Flores, Indonesia. Depth 15 m. Length 15 cm, female.

C

H. grayi. Maumere, Flores, Indonesia. Length 15 cm.

A

Starry Pipefish *Halicampus punctatus*

Yozia punctata Kamohara, 1952.
Mimase market, Kochi City, Japan.

Only known from sub-tropical Japanese waters, south to Okinawa. Occurs at moderate depths and most specimens are known from trawls. The specimens in the photographs were taken in 20–24 m depth on open sand with sparse reef nearby. Appears to be closely related to *Filicampus* and inclusion in present genus is doubtful. Length to 16 cm.

B

H. punctatus. Suruga Bay, Izu Peninsula. A 10 cm, B 12 cm. Hiroyuki UCHIYAMA.

Zavora Pipefish *Halicampus zavorensis*

Halicampus zavorensis Dawson, 1984.
Zavora, Mozambique.

Western Indian Ocean from Oman to off eastern South Africa. Algae reef and rubble habitat. Known from few specimens. Length to just over 10 cm.

H. zavorensis. Aliwal Shoal, South Africa. Length 102 mm. Phil HEEMSTRA.

Samoan Pipefish
Halicampus mataafae

Corythroichthys mataafae
Jordan & Seale, 1906. Samoa.

Widespread Indo-West Pacific. Many geographical variations and probably comprises a species-complex. Rock and coral habitats from tidal pools to 15 m depth. Length 10–14 cm depending on locality.

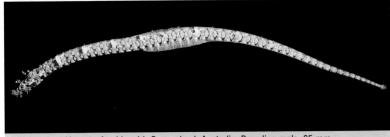

H. mataafae. Lizard I, Queensland, Australia. Brooding male, 65 mm.

Glittering Pipefish
Halicampus nitidus

Syngnathus nitidus Günther, 1873.
Australia.

Widespread West Pacific, northern Australia to southern Japan. Mainly found on coral reef flats to about 20 m depth. Easily recognised by colour but secretive and not often seen. A small species, largest known is 73 mm.

A

H. nitidus. Rabaul, PNG. Length 55 mm. Roger C. STEENE.

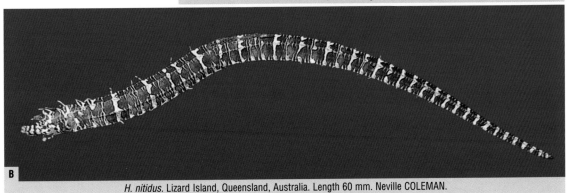

B

H. nitidus. Lizard Island, Queensland, Australia. Length 60 mm. Neville COLEMAN.

GENUS *APTERYGOCAMPUS* WEBER, 1913

Masculine. Type species: *Apterygocampus epinnulatus* Weber, 1913. A single tiny species recognised from the West Pacific. Thought to metamorphose from planktonic juveniles with major changes to fin arrangements.

Minute Pipefish *Apterygocampus epinnulatus*

Apterygocampus epinnulatus Weber, 1913.
Off Ceram, Indonesia.

West Pacific, known from few specimens. Most collected at night at the surface with lights, but holotype from 'reef'. This was a brooding male that lacked dorsal and pectoral fins, whilst its pouch-larvae possessed those fins. Planktonic specimens at similar size of adults have dorsal fin and pectoral fins. The female looks much like female *Acentronura*, but have a moderately large caudal fin. The pouch of the male holotype is sac-like, also similar to *Acentronura*. Because of minute size this species is easily overlooked. A few were found on shallow mud-flats with sparse seagrasses. Length to about 30 mm (largest known is 28 mm).

A. epinnulatus. 3 m depth, Maumere, Flores, Indonesia. Female, length 25 mm.

169

Masculine. Type species *Ichthyocampus belcheri* Kaup, 1856. Indo-West Pacific genus with at least 3 species. Small fishes from rock and coral reef habitats.

Most species are found on mixed algae-rubble habitats and probably do well in 'living rock' aquariums, but food will be a critical part as most feed on very small crustaceans.

Pale-blotched Pipefish
Phoxocampus diacanthus

Ichthyocampus diacanthus Schultz, 1943. Samoa.

Widespread West Pacific, ranging west to Sri Lanka and east to Samoa. Shallow protected rubble reefs to depths of 40 m. Length to 87 mm.

P. diacanthus. Lizard Island, Queensland, Australia. Length about 5 cm.

Black Rock Pipefish
Phoxocampus belcheri

Ichthyocampus belcheri Kaup, 1856. China.
Ichthyocampus nox Snyder, 1909. Okinawa, Japan. Presently regarded a synonym.

Widespread Indo-West Pacific, but Indian Ocean population has a barred colour pattern and is most likely to be a different species. Reported from tide pools to 15 m depth. Length to 80 mm.

P. belcheri (form *nox*). Kerama, Japan. Length about 40 mm. Atsushi ONO.

Trunk-barred Pipefish
Phoxocampus tetrophthalmus

Syngnathus tetrophthalmus. Bleeker, 1858. Kokos I, Indonesia.

Philippines and Indonesia, including western Irian Jaya. Lateral trunk ridge runs to at least 13th tail ring. This ends before 6th in other species, except for *P. kampeni.* Coral reef habitats to about 10 m depth. Length to 80 mm.

P. tetrophthalmus. Raja Ampat Islands, Irian Jaya. Length 65 mm. G.R. ALLEN.

Kampen's Pipefish
Phoxocampus kampeni

Ichthyocampus kampeni. Weber, 1913, Indonesia and New Guinea.

Indonesia and New Guinea. Previously included with *P. tetrophthalmus.* Mainly dark brown with thin black striations. Secretive on soft-bottom substrates to 20 m depth. Length to 80 mm.

P. kampeni. Port Moresby, PNG. Length 75 mm. Neville COLEMAN.

Masculine. Type species: *Siokunichthys herrei* Herald, 1953. Indo-Pacific genus with 5 small species. Several are known to live in association with soft or mushroom corals.

Mushroom-coral Pipefish
Siokunichthys nigrolineatus

Siokunichthys nigrolineatus Dawson, 1983. Moluccas Is, Indonesia.

Southern Indonesia and Papua New Guinea to Philippines. Lives amongst the tentacles of mushroom corals such as *Heliofungia actiniformes* in photographs. Sometimes the pipefish is mistaken by divers for a worm and the coral for an anemone. The pipefishes occur in small groups, usually of mixed sex and sizes. Depth range 10–20 m and habitat at bases of small drops of inner reefs, often adjacent to large lagoons which discharge large volumes of water over reefs on outgoing tides that usually carry plankton. Length to 80 mm.

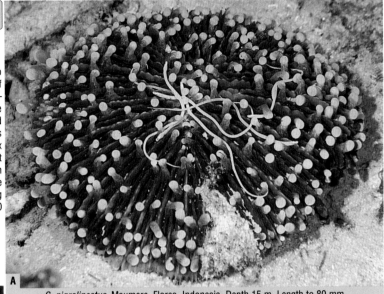

A

S. nigrolineatus. Maumere, Flores, Indonesia. Depth 15 m. Length to 80 mm.

B

S. nigrolineatus. Maumere, Flores, Indonesia.

C

S. nigrolineatus. Maumere, Flores, Indonesia. Depth 15 m. Length to 80 mm.

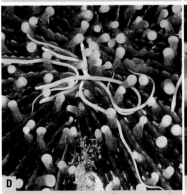

D

S. nigrolineatus. Maumere, Flores, Indonesia.

E

S. nigrolineatus. Milne Bay, Papua New Guinea. Depth 10 m. Length to 80 mm.

Soft-coral Pipefish *Siokunichthys breviceps*

Siokunichthys breviceps Smith, 1963.
Pinda, Mozambique.

Populations in West Indian Ocean and in West Pacific that probably represent two species. Life-colours of West Indian Ocean form is not known at this stage, except as uniform light cream for the type material. West Pacific form has a banded pattern. It is found in soft-coral habitats in coastal waters in depths to about 10 m. Length to 15 cm.

S. cf *breviceps.* Lizard Island, Queensland. Depth 7 m. Length 60 mm.

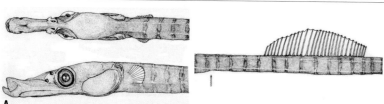

S. cf *breviceps.* Tulamben, Bali, Indonesia. Depth 6 m. Length 10 cm. Takamasa TONOZUKA.

Bentuvia's Pipefish *Siokunichthys bentuviai*

Siokunichthys bentuviai Clark, 1966.
Dahlak Archipelago, Red Sea.

A Red Sea endemic. Greenish with red bars on trunk. Lives in soft-coral (*Zenia* sp) habitats to 10 m depth. Length to 75 mm.

S. bentuviai. Red Sea. Brooding male, 59 mm. Various aspects of the head, pouch and trunk to tail section. After DAWSON.

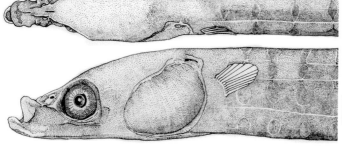
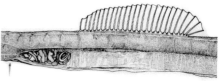

Herre's Pipefish
Siokunichthys herrei

Siokunichthys herrei Herald, 1953.
Mindanao I., Philippines.

Indonesia and Philippines and similar one in the northern Red Sea with fewer tail rings (undescribed). Since all know specimens were taken from planktonic samples, and none were adults, they may actually be pelagic stages of the other species of the area. Some morphological features may change between juvenile and adult stages, especially when transforming from pelagic to benthic life even at large planktonic stages. Largest sample ~78 mm long.

S. herrei. Fiji. Length 7 cm. Various aspects of the head and trunk to tail section. After DAWSON.

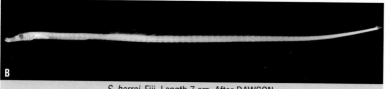

S. herrei. Fiji. Length 7 cm. After DAWSON.

Masculine. Type species: *Syngnathus deocata* Hamilton, 1822. A large group of mainly freshwater pipefishes with at least 20 species, divided over the entire Indo-Pacific and 2 in Atlantic seas. Dividable into 5 subgenera, based on variations of ridges. Adults of all species occur most commonly in coastal streams, rivers or lakes with low salinity, in which they breed. Young and subadults are often found in coastal waters and may disperse over large coastal regions during wet-seasons. All species have a broad geographical distribution in the various regions, but their identity is primarily based on preserved specimens and some may comprise additional taxa. It is unlikely that a species from the West Pacific is the same as one in the West Indian Ocean. Localised colour forms are to be expected. Males have a long brood pouch that runs along almost the entire ventral trunk.

Flat-nose River Pipefish
Microphis biocellatus

Coelonotus biocellatus Günther, 1870. East-Indian Archipelago?.

Southern Indonesia to Japan and Papua New Guinea to Fiji. Similar species *M. argulus*, western Indian Ocean, and *M. platyrhynchus*, east Pacific. Snout short, and has double series of small ocelli along trunk, spaced equal in distance between series. Occurs in freshwater streams and rivers. Length to 15 cm.

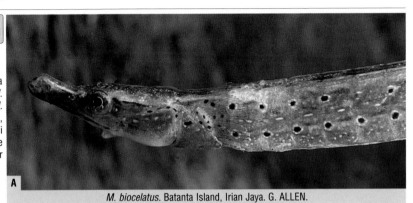

M. biocelatus. Batanta Island, Irian Jaya. G. ALLEN.

M. biocelatus. Yutsun Riv., Iriomote I, Japan. T. ZUZUKI.

Long-fin River Pipefish
Microphis mento

Syngnathus mento Bleeker, 1856. Manado, Sulawesi, Indonesia.

Northern Indonesia, Philippines and Papua New Guinea. Similar *M. fluviatilis* in western Indian Ocean. Head grey, dark line through the eye, trunk with several yellow-brown longitudinal lines. Occurs in freshwater streams and rivers. Length to 15 cm.

M. mento. Batanta Island, Irian Jaya. G. ALLEN.

Synonyms: *Hemithylacus rocaberti* Duméril, 1870, and *Parabelonichthys kellersi* Fowler, 1943.

Stream Pipefish
Microphis brevidorsalis

Doryrhamphus brevidorsalis de Beaufort, 1913. Buru, Indonesia.

Indonesia and into Pacific, to Solomon Islands and Fiji. Short-snouted. Trunk with series of large, vertically-elongated button-like spots, one on each trunk ring. Lives in freshwater streams and rivers. Length to 14 cm.

M. brevidorsalis. Male. Gam River, Misool, Irian Jaya. G. & M. ALLEN.

Ragged-tail Pipefish *Microphis retzii*

Syngnathus retzii Bleeker, 1856.
Manado, Sulawesi, Indonesia.

Widespread from Indonesia to Southern Japan, Philippines and Papua New Guinea to central Pacific. Pelagic young have elongated rays in centre of caudal fin, but these maybe present in the other young members in the genus. Occurs in freshwater streams and rivers, sometimes found far inland. Length to 15 cm.

Synonyms: *Microphis caudatus* Peters, 1869, *Microphis torrentius* Jordan & Seale, 1906, and *Doryichthys retzii albidorsum* Fowler, 1944.

A

M. retzii. Miero River, Madang, Papua New Guinea. Jerry ALLEN.

B

M. retzii. Adola River, Misool, Irian Jaya. Jerry & Mark ALLEN.

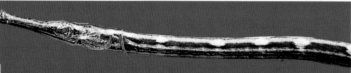

C

M. retzii. Uranchi River, Iriomote I, Japan. Toshiyuki SUZUKI.

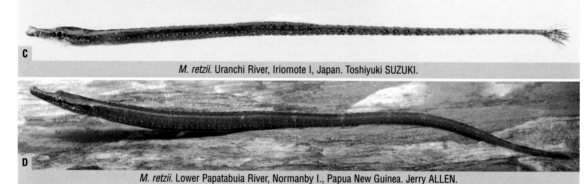

D

M. retzii. Lower Papatabuia River, Normanby I., Papua New Guinea. Jerry ALLEN.

Manado River Pipefish
Microphis manadensis

Syngnathus manadensis Bleeker, 1856.
Manado, Indonesia.

Northern Indonesia to Palau and Solomon Islands. Females distinctive with evenly spaced series of pale saddles over the back. Occurs in freshwater streams and rivers, sometimes far inland. Length to 15 cm.

Synonyms: *Doryichthys bernsteini* Bleeker, 1867.

A

M. manadensis. Female. Marur River, Western Waigeo, Irian Jaya. G. & M. ALLEN.

B

M. manadensis. Male. Marur River, Western Waigeo, Irian Jaya. G. & M. ALLEN.

Bud River Pipefish *Microphis leiaspis*

Syngnathus leiaspis Bleeker, 1853.
Manado, Indonesia.

West Pacific, Iriomote, Japan and Philippines to Indonesia and east to Fiji. Closely related species in Japan (*M. yoshi*) and Western Indian Ocean (*M. vaillantii*). Occurs in lower reaches of rivers and in brackish estuaries. Reaches 20 cm.

Synonyms: *Syngnathus budi* Bleeker, 1856.

A

M. leiaspis. Adola River, Misool, Irian Jaya. G. & M. ALLEN.

B

M. leiaspis. Yutsun River, Iriomote I, Japan. Toshiyuki SUZUKI.

Bar-head River Pipefish
Microphis yoshi

Siphostoma yoshi Snyder, 1909.
Tanegashima, Japan.

Only known from the type locality and Ryukyu Islands. Sometimes regarded as a variation of *M. leiaspis*.

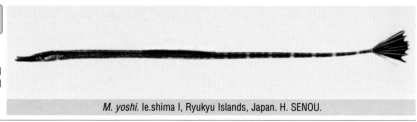

M. yoshi. Ie.shima I, Ryukyu Islands, Japan. H. SENOU.

Thousand-spot River Pipefish
Microphis millepunctatus

Doryichthys millepunctatus Kaup, 1856. Réunion, Madagascar.

Appears to be widespread western Indian Ocean. Similar species *M. brachyurus* in West Pacific. Lives primarily in brackish estuaries and lower reaches of freshwater streams. Length to 16 cm.

M. millepunctatus. Mauritius. Length 10 cm. Phil HEEMSTRA.

Flowing River Pipefish
Microphis fluviatilis

Syngnathus fluviatilis Peters, 1852. Mozambique.

Western Indian Ocean region, Kenya to South Africa and Madagascar. Length to 23 cm.

Synonyms: *Syngnathus zambezensis* Peters, 1855, and *Belonichthys sanctipauli* Sauvage, 1879.

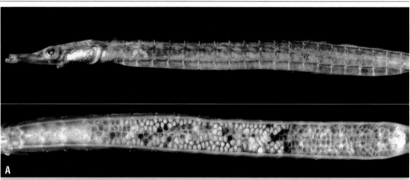

A

M. fluviatilis. Madagascar. Brooding male, length 15 cm. After DAWSON.

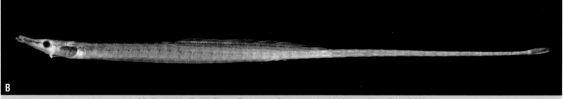

B

M. fluviatilis. Rufiji River, Tanzania. Female, length 16 cm. After DAWSON.

Spinach River Pipefish
Microphis spinachioides

Doryichthys spinachioides Duncker, 1915.
Sepik River, New Guinea

Only known from the type locality. Length to 15 cm.

M. spinachioides. Sepik River, PNG. Juvenile, length 8 cm. After DAWSON.

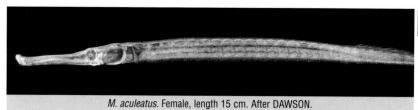

Atlantic River Pipefish
Microphis aculeatus

Doryichthys aculeatus Kaup, 1853.
West African coast, East Atlantic.

Similar to Pacific *Microphis brachyurus,* and *M. lineatus* from western Atlantic. Length to 18 cm.

M. aculeatus. Female, length 15 cm. After DAWSON.

Tawarin River Pipefish
Microphis caudocarinatus

Doryichthys caudocarinatus
Weber, 1908.
Tawarin River, northern Irian Jaya.

Only known from a single juvenile specimen, 70 mm long

M. caudocarinatus. Tawarin River, Northern Irian Jaya. Holotype, length 7 cm. After DAWSON.

Dumbea River Pipefish
Microphis cruentus

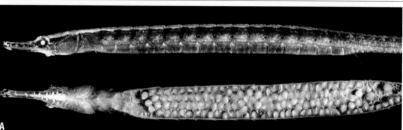

Microphis cruentus
Dawson & Fourmanoir, 1981.
Dumbéa River. New Caledonia.

Endemic to New Caledonia. Length to about 16 cm.

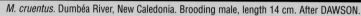

A

M. cruentus. Dumbéa River, New Caledonia. Brooding male, length 14 cm. After DAWSON.

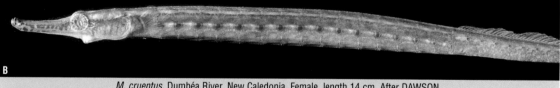
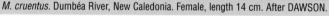

B

M. cruentus. Dumbéa River, New Caledonia. Female, length 14 cm. After DAWSON.

Crocodile-tooth River Pipefish
Microphis cuncalus

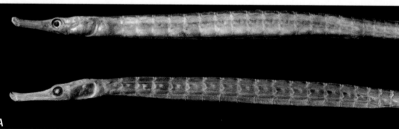

Syngnathus cuncalus
Hamilton Buchanan, 1822. India.

India, Bangladesh and Sri Lanka. Rivers and estuaries. Length to about 18 cm.
Synonyms: *Paramicrophis schmidti* Klausewitz, 1855, and *Doryichthys chokderi* Rahman, 1976.

A

M. cuncalus. Ganges and Hoogly Rivers, India. Female above, male below, length 14 cm. After DAWSON.

B

M. cuncalus. India. Juvenile, length 12 cm. After DAWSON.

Indian Inland Pipefish
Microphis deocata

Syngnathus deocata
Hamilton Buchanan, 1822. Rivers of northern Bengal and Bahar, India.

Only known from the India region. Female apparently colourful on trunk with distinctive upside-down 'Y'-shaped markings. Length to 14 cm.

A

M. deocata. Phura River, Assam, India. Brooding male, length 12 cm. After DAWSON.

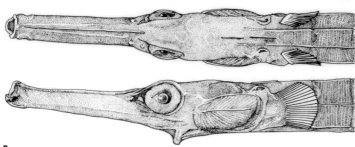

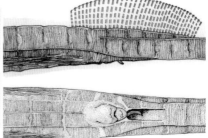

B

M. deocata. Phura River, Assam, India. Female, length 12 cm. Various aspects of the head and trunk to tail section. After DAWSON.

Myanmar Inland Pipefish
Microphis dunckeri

Doryichthys dunckeri
Prashad & Mukerji, 1929.
Upper Myanmar (Burma).

Only known from the upper Irrawaddy River drainage, some 1500 km or more from the sea.

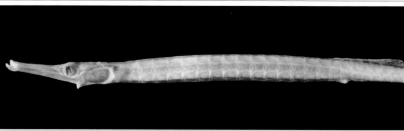

M. dunckeri. Irrawaddy River, Myanmar. Female, 12 cm. After DAWSON.

Andaman River Pipefish
Microphis insularis

Doryichthys insularis Hora, 1825.
South Andaman Island.

Endemic to the Andaman Islands.
Length to about 16 cm.

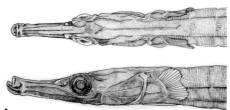

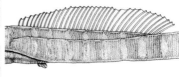

A

M. insularis. Various aspects of the head and trunk to tail section. After DAWSON.

B

M. insularis. South Andaman Island. Female, length 13 cm. After DAWSON.

Philippine River Pipefish *Microphis jagorii*

Microphis jagorii Peters, 1869. Philippines.

Only known from the Philippines and southern-most Japan islands but very similar to *M. manadensis*. Length to about 18 cm.

A

M. jagorii. Philippines. Central section of male, length 14 cm. After DAWSON.

B

M. jagorii. Nakara River, Iriomote I, Japan. Toshiyuki SUZUKI.

Short-tail River Pipefish *Microphis brachyurus*

Syngnathus brachyurus Bleeker, 1853. Java & Sumatra, Indonesia.

Widespread West Pacific, ranging into Indian Ocean to Sri Lanka, and into central Pacific. Similar to *M. millepunctatus* in the western Indian Ocean. Long-snouted, usually snout dark with pale barring or spots forming barred pattern.

Synonyms:
Syngnathus polyacanthus Bleeker, 1856.
Doryichthys hasselti Kaup, 1856.
Doryichthys auronitens Kaup, 1856.
Microphis bleekeri Day, 1865.
Microphis jouani Duméril, 1870.
Doryichthys philippinus Fowler, 1919.

M. brachyurus. Nishihonera River, Iriomote I, Japan. Toshiyuki SUZUKI.

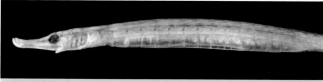

M. ocellatus. Rumbia, Kalimantan, Indonesia. Female, l. 10 cm. After DAWSON.

Sri Lanka River Pipefish *Microphis ocellatus*

Doryichthys ocellatus Duncker, 1910. Sri Lanka.

Restricted to coastal areas of Andaman Sea, Sri Lanka to northern Sumatra and Kalimantan. Length to about 12 cm.

M. pleurostictus. Luzon Island, Philippines. Female, 12 cm. After DAWSON.

Luzon River Pipefish *Microphis pleurostictus*

Doryichthys pleurostictus Peters, 1869. Luzon Island, Philippines.

Only known from type material, from Lake Bato and Yassot Creek.

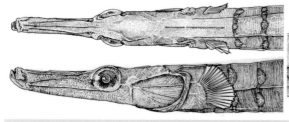
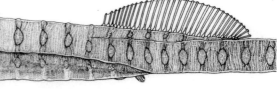

M. pleurostictus. Luzon Island, Philippines. Female, 12 cm. Various aspects of the head and trunk to tail section. After DAWSON.

Other valid species of *Microphis*.

Opossum River Pipefish *Microphis lineatus*

Doryichthys lineatus Kaup, 1856. Veracruz, Mexico.
West Atlantic, Mexico to Brazil.

Madagascar River Pipefish *Microphis vaillantii*

Coelonotus vaillantii Juillerat, 1880. Madagascar.
Only known from Madagascar. Similar to the Pacific *M. leiaspis*

Masculine. Type species: *Hippichthys heptagonus* Bleeker, 1849. Widespread Indo-West Pacific genus with at least 6 species, some of which are wide-ranging, and an additional 20 or so descriptions are thought to be junior synonyms. They are primarily coastal estuarine species that live in brackish waters and some in freshwater.

Madura Pipefish *Hippichthys heptagonus*

Hippichthys heptagonus Bleeker, 1849. Madura, Indonesia.

Widespread Indo-West Pacific from east African coast to Solomon Island, and tropical Japan to tropical eastern Australia. Probably comprises a species complex. Lives in the lower reaches of streams and estuaries, sometimes in lakes. Length to 15 cm.

Synonyms: *see list below*

A

H. heptagonus. Female. Aird Hills, Papua New Guinea. G.R. ALLEN.

B

H. heptagonus. Female. Urauchi River, Iriomote I. Japan. H. SENOU.

C

H. heptagonus. Female. Aird Hills, Papua New Guinea. G.R. ALLEN.

D

H. heptagonus. Brooding male. Johnston River, Innisfail, Queensland. From Mangroves, brackish. Bruce COWELL.

Synonyms:
Syngnathus djarong Bleeker, 1849. Java, Indonesia.
Syngnathus helfrichii Bleeker, 1855. Borneo, Indonesia.
Syngnathus parviceps Ramsay & Ogilby, 1887. Clarence River, New South Wales, Australia.
Corythroichthys pullus Smith & Seale, 1906. Mindanao, Philippines.
Corythroichthys matterni Fowler, 1918. Philippines.
Bombania luzonica Herre, 1927. Lake Taal, Luzon, Philippines.
Bombania uxorius Herre, 1935. Waigu, Indonesia.
Syngnathus djarong luzonica Aurich, 1935. Lake Taal, Luzon, Philippines.

Belly-barred Pipefish *Hippichthys spicifer*

Hippichthys spicifer Rüppell, 1838. Tor, Red Sea.

Indo-West Pacific. Red Sea and east African coast to Samoa, tropical Japan and to north-eastern Australia, if representing a single species. Shallow coastal estuaries, mangroves, and lower reaches of rivers. Some geographical variations that need further investigation. Length to 15 cm.

Synonyms: *Syngnathus hunni* Bleeker, 1860. Sumatra, Indonesia; *Syngnathus gracilis* Steindachner, 1903. Ternate, Indonesia; *Micrognathus suvensis* Herre, 1953. Suva, Fiji.

A

H. spicifer. Brooding male. Daintree River, Qld. Roger C. STEENE.

B

H. spicifer. Female. Johnston River, Innesfall, Queensland. From mangroves, estuarine. Bruce COWELL.

C

H. spicifer. Brooding male. Honera River, Iriomote I., Japan. H. SENOU.

Beady Pipefish *Hippichthys penicillus*

Syngnathus penicillus Cantor, 1849. Sea of Pinang, Malaysia.

Widespread West Pacific, ranging from tropical Japan to northern half of Australia, south to Moreton Bay. Sheltered estuaries in mangroves and *Zostera* seagrasses, entering lower reaches of streams and rivers affected by tides. Length to 16 cm.

Synonyms: *Syngnathus argyrostictus* Kaup, 1856. Java, Indonesia; *Syngnathus biserialis* Kaup, 1856. India; *Syngnathus altirostris* Ogilby, 1890. Moreton Bay, Queensland, Australia; *Corythroichthys quinquarius* Snyder, 1911. Tanegashima, Japan; *Hippichthys gazella* Whitley, 1947. Broome, Western Australia.

A

H. penicillus. Boggy Creek, Moreton Bay, Qld. Bruce COWELL.

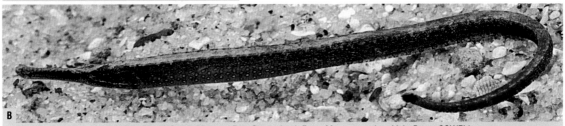

B

H. penicillus. Boggy Creek, Moreton Bay, Queensland. From Mangroves, brackish. Bruce COWELL.

C

H. penicillus. Kamo River, Kii Peninsula, Japan. H. SENOU.

Blue-speckled Pipefish *Hippichthys cyanospilos*

Syngnathus cyanospilos Bleeker, 1854. Banda Neira, Indonesia.

Widespread Indo-West Pacific from Red Sea and east African coast to Fiji, Philippines and to north-eastern Australia. Shallow coastal tidal estuaries, mangroves, and brackish water. Variable pale yellow to nearly black. Dorsal fin with series of small black spots, diagnostic for this species. Length to 16 cm.

Synonyms:
Syngnathus mossambicus Peters, 1855. Mozambique.
Syngnathus kuhlii Kaup, 1856. Indonesia.
Doryichthys spaniaspis Jordan & Seale, 1907. Philippines.
Parasyngnathus wardi Whitley, 1948. Queensland, Australia

A

B

H. cyanospilos. Female. Gilimanuk, Bali, Indonesia. Roger C. STEENE (**A-B**).

C

H. cyanospilos. Male. Gilimanuk, Bali, Indonesia. Akira OGAWA.

D

H. cyanospilos. Brooding male. Gilimanuk, Bali, Indonesia. Takamasa TONOZUKA.

E

H. cyanospilos. Female. Ishigaki I., Japan. H. SENOU.

181

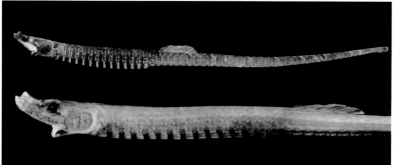

Short-keel Pipefish
Hippichthys parvicarinatus

Syngnathus parvicarinatus Dawson, 1978. Darwin, Australia

Only known from Kimberleys to Cape York, Northern Australia. In estuarine and freshwater habitats. Length to about 12 cm.

H. parvicarinatus. Darwin Australia. Female, 84 mm above, and male, 81 mm below. After DAWSON.

GENUS *ICHTHYOCAMPUS* KAUP, 1853

Masculine. Type species: *Syngnathus carce* Hamilton Buchanan, 1822. Monotypic, freshwater southeast Asia.

Indian Freshwater Pipefish
Ichthyocampus carce

Syngnathus carce Hamilton Buchanan, 1822. Ganges River, 'tide-ways', India.

Coastal regions of India, Sri Lanka to western Indonesia to Sulawesi. Usually found in streams, rivers and estuaries. Plain brownish with diffused barring on trunk, but ventral trunk turns bright red during courtship (as in photographs). Length to 12 cm.

An attractive aquarium species that readily breeds in freshwater.

I. carce. **A-B** Courting pair, upper in **B** is male. **C** Hatchlings. Aquarium Germany, probably Sri Lanka import. Uwe WERNER.

Masculine. Type species: *Doryichthys bilineatus* Kaup, 1856 (= *Syngnathus deokhatoides* Bleeker, 1853). West Pacific genus with 4 species. They are coastal species that live primarily in freshwater habitats.

Additional photographs are needed, especially from different areas. Contributions and additional information will be appreciated and credited.

Asian River Pipefish
Doryichthys boaja

Syngnathus boaja Bleeker, 1851.
Kalimantan, Indonesia.

Coastal regions of southeast Asia from Thailand to Vietnam, and the large Indonesian Islands, east to Sulawesi. Mainly found in streams and rivers. Adults distinctly marked with barred pattern on trunk. The largest freshwater pipefish, reaching at least 44 cm in length.

An attractive aquarium species that is often on display in public aquariums.

D. boaja. Aquarium, Germany.

D. boaja. Aquarium. Frankfurt Zoo, Germany. Lengths to about 25 cm.

Large-spots River Pipefish
Doryichthys deokhatoides

Syngnathus deokhatoides
Bleeker, 1853.
Sumatra and Kalimantan, Indonesia.

Kalimantan and Sumatra, Indonesia and Malay Peninsula. Freshwater habitats. Length to 17 cm.

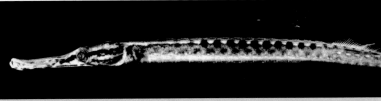

D. deokhatoides. Thailand. Female, length 10 cm. After DAWSON.

Sambas River Pipefish
Doryichthys heterosoma

Syngnathus heterosoma Bleeker, 1851.
Kalimantan, Indonesia.

Sambas River, Kalimantan and the Natuna Islands. Length to 35 cm.

D. heterosoma. Sambas River, Kalimantan. Female, length about 30 cm. After DAWSON.

Martens' River Pipefish
Doryichthys martensii

Syngnathus martensii Peters, 1869.
Kalimantan, Indonesia.

Kalimantan and Sumatra, Indonesia and the Malay Peninsula ranging west to Thailand. Freshwater habitats. Length to 17 cm.

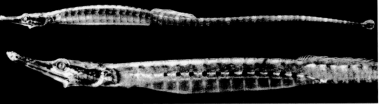

D. martensii. Kalimantan, female 10 cm above, Thailand, male 11 cm below. After DAWSON.

183

Masculine. Type species: *Urocampus nanus* Günther, 1870. West Pacific genus with 2 species recognised, divided between Japan and Australia. They are coastal species that live primarily in marine estuarine habitats. Young are non-pelagic and species are probably more localised than previously thought. Adults may travel with loose weed on the bottom. Several additional species are to be expected.

A **B**

U. carinirostris. Courting pair, gravid female left. Wallaga Lake, New South Wales.

Hairy Pipefish *Urocampus carinirostris*

Urocampus carinirostris Castelnau, 1872.
Melbourne, Australia.

Common in marine estuaries of New South Wales, ranging into southern Queensland, Tasmania, and scattered population along the south coast, west to the Perth region. Populations in Tasmania mainly in brackish estuaries. Also reported from Papua New Guinea and Japan. No doubt represents several additional species, as no reef fish shares a distribution range between temperate Tasmania and tropical Papua New Guinea. Usually common where found, inhabiting *Zostera*, eelgrass habitats, or on sheltered algae-rubbly mixed sparse reefs to about 5 m depth. Hatchlings are non-pelagic and settle in vicinity of parents. Length to 10 cm.

Urocampus coelorhynchus Günther, 1873, Sydney, and *U. guentheri* Duncker, 1909, Shark Bay, W.A. were thought to be synonyms but need to be investigated.

C **D** **E** **F**

U. carinirostris. **C** spawning pair, **D** pregnant male. Wallaga Lake, N.S.W.   *U. carinirostris.* Birthing male. Wallaga Lake, N.S.W.

G

U. carinirostris. Hatchling. Wallaga Lake, N.S.W.

H

U. carinirostris (U. coelorhynchus). Manly, Sydney Harbour. D. 4 m. L. 60 mm.

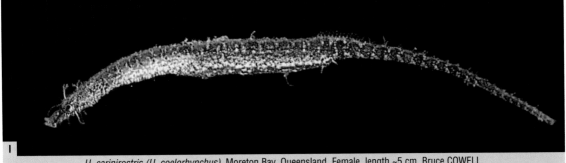

U. carinirostris (U. coelorhynchus). Moreton Bay, Queensland. Female, length ~5 cm. Bruce COWELL.

U. carinirostris (U. guentheri). Shark Bay, Western Australia. Brooding male, length ~5 cm. Clay BRYCE.

Barbed Pipefish *Urocampus nanus*

Urocampus nanus Günther, 1870. Manchuraria, Japan.
Urocampus rikuzenius Jordan & Snyder, 1901. Matsushima Bay, Japan.

Mainly known from Honshu, Japan. Estuarine species, usually amongst *Zostera* seagrasses. Length to 150 mm.

U. nanus. Various aspects of the head and trunk to tail section. After DAWSON.

U. nanus. Lake Hamana, Shizuoka Pref. Japan. H. Senou.

Masculine. Type species: *Syngnathus acus* Linnaeus, 1758. Large genus with 32 species as presently defined, variously distributed in Atlantic and Pacific Oceans, but poorly represented in the Indo-West Pacific. However, a number of species are provisionally included and may belong in other genera.

Additional photographs are needed. Contributions and additional information will be appreciated and credited.

A

S. acus. Baltic Sea. Tony HOLM.

Great Pipefish *Syngnathus acus*

Syngnathus acus Linnaeus, 1758. Europe.

North-east Atlantic, south to islands off the north-east African region, and Mediterranean. A large, long-snouted species with a somewhat humped head as shown in **B**. One of the first described pipefishes and it represents a form that closely resembles earlier ancestral that gave rise to various forms in the Indo-West Pacific that evolved at different and faster rates into other genera. *S. acus* occurs commonly in relatively shallow and sheltered habitats with seagrasses and algaes. Usually seen in depths less than 15 m, but has been trawled in much deeper water on open rubble substrates (to 90 m). May enter brackish waters. Breeds in summer, incubating about 400 or more eggs over 5 weeks, that is comparable with similar sub-tropical species in the southern hemisphere. Length to 45 cm.

S. rubescens Risso, 1810, from Nice, Mediterranean is regarded as a synonym.

Similar species in South Africa and West Indian Ocean, *S. temminckii* Kaup, 1856, Cape of Good Hope, and its synonyms *S. brachyrhynchus*, from Réunion, *S. delalandi*, from Cape of Good Hope, both by Kaup, 1856, and *S. alrenans* Günther, 1870, from the Seychelles.

B

S. acus. Mediterranean. Phil WOODHEAD.

C

S. acus. Cornwall, England. Charles HOOD.

Dark-flank Pipefish *Syngnathus taenionotus*

Syngnathus taenionotus Castrini, 1871. Venice, Italy.

Only known from the Adriatic and Black Seas, but probably widespread along the coastal parts of the north-east Mediterranean. Males with dark stripe along snout and trunk. Female mainly with stripe along snout. Grey to yellow, latter usually female. Shallow inshore seagrass beds. Length to 20 cm.

It appears that in the Black Sea, males outnumber females in Syngnathidae.

S. taenionotus. **B** Female. **A** & **C** Males congregating during the spawning time. Bulgarian south coast, Black Sea. Werner FIEDLER.

S. taenionotus. Pregnant male. Golfe du Lion, Mediterranean Sea, France. Patrick LOUISY.

S. taenionotus. Female. Golfe du Lion, Mediterranean Sea. Patrick LOUISY.

Syngnathus rostellatus Nilsson, 1855. Sweden.

Subtemperate North Seas, ranging south to Gulf of Biscay. Maybe confused with sub-adult *S. acus* but has much shorter snout (about half of total head length). Found from estuaries to open sandy substrates to about 20 m depth, but most common in less than 2 m in weeds. Often in loose weeds. Length to 20 cm.

S. rostellatus. Female. Baltic Sea, near Wismar. Werner FIEDLER.

S. rostellatus. Brooding males. Baltic Sea, near Wismar. Werner FIEDLER.

S. rostellatus. Golfe du Lion, Mediterranean Sea. Patrick LOUISY.

S. rostellatus. Golfe du Lion, Mediterranean Sea. Patrick LOUISY.

Syngnathus abaster Risso, 1826.
Probably Nice, France.

Widespread Mediterranean and Black Seas, ranging north along Atlantic coast to Biscay. Mainly in shallow estuarine habitats in depths less than 5 m. Variable brown to green with dark or pale spots or bars on trunk and tail. Often has dark markings along base of dorsal fin. Snout short compared to similar species in same area, less than half of head length. Length to 15 cm, but grows larger in Black Sea, to 21 cm.

S. abaster. Courting pair. Golfe du Lion, Mediterranean Sea. Patrick LOUISY.

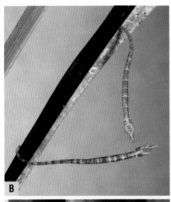

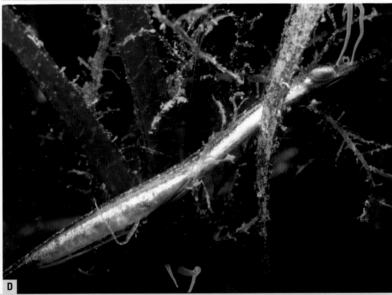

S. abaster. **B** & **C** hatchlings. **D** male, giving birth. Golfe du Lion, Mediterranean Sea. Patrick LOUISY.

S. abaster. Female. Golfe du Lion, Mediterranean Sea. Patrick LOUISY.

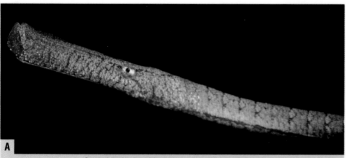

Syngnathus typhle Linnaeus, 1758. Europe.

North-east Atlantic, ranging to Baltic seas and Mediterranean. Recognised be snout as deep as height of body. Variable brown to green with habitat. In vegetated reef habitats to about 20 m depth, usually amongst *Zostera* grasses or weeds. Length to 35 cm.

Remarks: two forms or subspecies are recognised. *Syngnathus typhle typhle* in the Atlantic, and *Syngnathus typhle rondeletii* in the Mediterranean.

S. typhle. Malta, Mediterranean. Victor FABRI.

S. typhle. Majorque, Baléares. Patrick LOUISY.

S. typhle (typhle). Italy, Mediterranean. Thomas PAULUS.

S. typhle (rondeletii). Atlantic low-snout form, subspecies *typhle*. Frankfurt Zoo, Germany.

Delicate-nose Pipefish
Syngnathus tenuirostris

Syngnathus tenuirostris Rathke, 1837.
Ukraine, Black sea.

Mainly known from Black Sea, southern Adriatic Sea and Sea of Azov, but also in scattered localities of the northern Mediterranean, wet to France. Occurs inshore algae-weed rubble or reefs from the shallows to about 25 m. Colour dark-grey to reddish brown or yellow with diffused broad banding over trunk and tail. Dorsal fin with dark spots. Snout long and fairly straight. Length to almost 40 cm.

A

S. tenuirostris. Golfe du Lion, Mediterranean Sea. Patrick LOUISY.

B

S. tenuirostris. Golfe du Lion, Mediterranean Sea. Patrick LOUISY.

C

S. tenuirostris. Male. Golfe du Lion, Mediterranean Sea. Patrick LOUISY.

Variegated Pipefish *Syngnathus variegatus*

Syngnathus variegatus Pallas, 1814.
Ukraine, Black Sea.

Little known species, possibly endemic to the Black Sea.

A

B

S. variegatus? Black Sea, Bulgarian coast. Werner FIEDLER.

Sargassum Pipefish
Syngnathus pelagicus

Syngnathus pelagicus Linnaeus, 1758.
Originally described from drifting rafts of sargassum weeds in Atlantic.

Widespread sub-tropical western Atlantic waters, north of equator. Reports from elsewhere are probably based on other species. Usually found with *Sargassum* seaweed rafts in open oceanic waters. Probably one of the least evolved species. Length to 17 cm.

S. pelagicus. Western Atlantic. Paul HUMANN.

Seaweed Pipefish
Syngnathus schlegeli

Syngnathus schlegeli Kaup, 1856.
Japan, China.
Syngnathus acusimilis Günther, 1873.
China

Sub-tropical, China seas and southern Japan. Harbours, estuaries and sheltered coastal bays. Usually in *Zostera* seagrasses or algal reefs to about 15 m depth. Length to 30 cm.

S. schlegeli. Izu Peninsula, Japan. Hiroyuki UCHIYAMA.

African Long-nose Pipefish
Syngnathus temminckii

Syngnathus temminckii Kaup, 1856.
Cape of Good Hope.

Southern water of South Africa. Occurs in shallow bays and estuaries, as well as offshore where reported to over 100 m depth. Found on low reefs with algae and sessile invertebrates or amongst *Zostera* seagrasses. Basic colour varies to suit habitat, ranging from yellow, brown or green with spotting or faint banding. Previously included with *Syngnathus acus* that has similar meristics, but differs considerably in shape of the head and general colour, and may reach a maximum length of 45 cm. Adult *S. temmickii* are usually about 20 cm, reported to 30 cm.

A

B

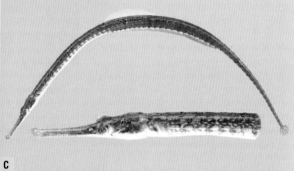

C

S. temminckii. S. Africa. **A** female. **B** brooding male. L. 20 cm. Guido ZSILAVECZ. **C** Kariega Riv. mouth. Male, 164 mm. Phil HEEMSTRA.

Follett's Pipefish *Syngnathus folletti*

Syngnathus folletti Herald, 1942. Flores I., Montevideo, Uruguay.

West Atlantic. Known as **Southern Pipefish** in the U.S.A. Has 16 trunk rings, 35–38 tail rings. Head 9.3 in SL. Dorsal-fin base-length 0.8 in head-length. Eggs about 100. South American coast, southern Brazil and northern Argentina. Algae and rock bottom to 30 m. Also in estuaries. Length to 20 cm.

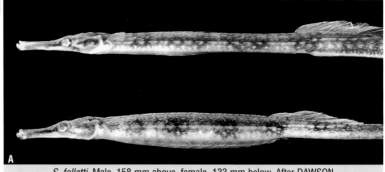

S. folletti. Male, 158 mm above, female, 133 mm below. After DAWSON.

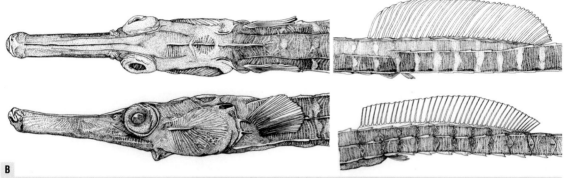

S. folletti. Various aspects of the head and trunk to tail section. After DAWSON.

Dusky Pipefish *Syngnathus fuscus*

Syngnathus fuscus Storer, 1839. Nahant, Massachusetts, U.S.A.

West Atlantic, Canadian and American coasts south to Florida. 19 trunk rings, 53–58 tail rings Head length 8 in SL. Snout 5 in length. 100–800 eggs. !0 days incubation. To 50 m depth. L. to 30 cm.

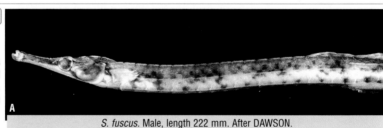

S. fuscus. Male, length 222 mm. After DAWSON.

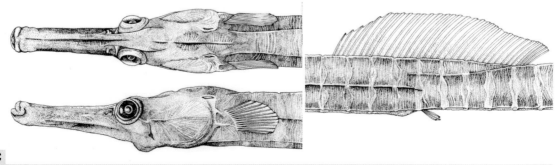

S. fuscus. Juvenile, length 110 mm. After DAWSON.

S. fuscus. Various aspects of the head and trunk to tail section. Female, length 177 mm. After DAWSON.

Chain Pipefish *Syngnathus louisianae*

Syngnathus louisianae Günther, 1870.
New Orleans, Louisiana, U.S.A.

West Atlantic. 19-21 trunkrings. Snout 10 in length 33 dorsal fin rays. Snout 1.7 in SL. Length to 35 cm. Brood 450–900 eggs.

S. louisianae. Female, length 170 mm upper (+drawing), male, 265 mm below. After DAWSON.

Hurghada Pipefish
Syngnathus macrophthalmus

Syngnathus macrophthalmus Duncker, 1915.
Suez, Red Sea.

Only known from northern Red Sea, south to Hurghada. Length to 12 cm.

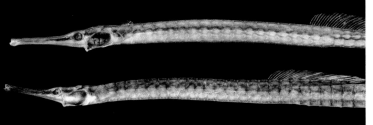

S. macrophthalmus. Suez. Juvenile, length 65 mm, paratype. After DAWSON.

Island Pipefish *Syngnathus insulae*

Syngnathus insulae Fritzsche, 1980.
Caleta Malpómene, Isla Guadaloupe, México.

Endemic to Isla Guadaloupe, México, but possibly a variety of *S. leptorhynchus*. Found on algae reef in 20-35 m depth, but sometimes in sargassum rafts. Length to 21 cm.

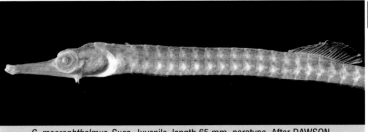

S. insulae. Baja California, México. Subadult, length 12 cm, paratype. After DAWSON.

Bay Pipefish *Syngnathus leptorhynchus*

Syngnathus leptorhynchus Girard, 1854.
San Diego, California.

Eastern Pacific, southern Alaska to Baja California, México. Near identical to *S. schlegeli* & *S. acus*. Some north-south clinal variations, typically with higher meristic values in cooler zones. Occurs amongst *Zostera* eel grasses in sheltered bays and estuaries. Length to 30 cm.

S. leptorhynchus. Puget Sound, U.S.A. Female, length 176 mm. After DAWSON.

Weed Pipefish
Syngnathus macrobrachium

Syngnathus macrobrachium Fritzsche, 1980.
Huasco, Chile.

Eastern Pacific, Peru and Chile. Length to 22 cm.

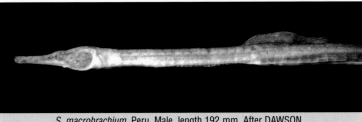

S. macrobrachium. Peru. Male, length 192 mm. After DAWSON.

Makax's Pipefish *Syngnathus makaxi*

Syngnathus scovelli makaxi Herald & Dawson, 1972
Laguna Makax, Isla Mujeres, México.

Only known from type-locality and nearby coast.
Lives in warm sub-tidal zones in dense algal or sea-grass habitats. Length to 70 mm.
Trunk 14. Dorsal 22–27. Short pectorals, 6.7 in head.

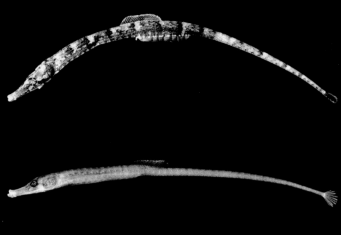

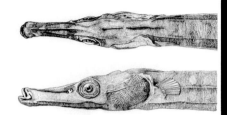

S. makaxi. Left: female, 56 mm. Right: Male, 45 mm above, female, 42 mm below. All are paratypes. After DAWSON.

Bull Pipefish *Syngnathus springeri*

Syngnathus springeri Herald, 1942.
Lemon Bay, Englewood, Florida.

West Atlantic. Western Bahamas, northern Gulf of Mexico and south to the Dry Tortugas. In 18 to 100+ m depth. Young in floating Sargassum. Adults often caught mid-water. Length to 35 cm.
Brood with up to 1400 eggs.

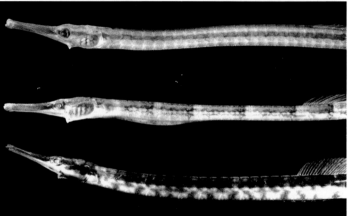

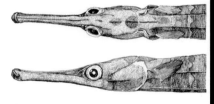

S. springeri. Left: female, 271 mm. Right: Juveniles, 135 mm above & 104 mm centre, male, 284 mm below. After DAWSON.

Barred Pipefish *Syngnathus auliscus*

Siphostoma auliscus Swain, 1882.
San Diego, Santa Barbara, California.

Eastern Pacific from California to Peru. Bays and estuaries amongst weeds. Sometimes in floating sargassum weeds. Length to about 18 cm.

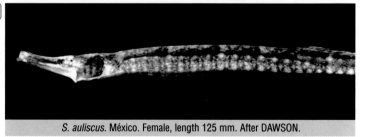

S. auliscus. México. Female, length 125 mm. After DAWSON.

Kelp Pipefish *Syngnathus californiensis*

Syngnathus californiensis Storer, 1845.
California.

Eastern Pacific, California region. Associates with *Macrocystis* kelp. Length to 20 cm.

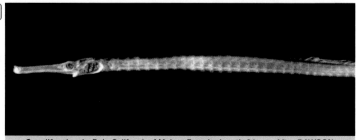

S. californiensis. Baja California, México. Female, length 21 cm. After DAWSON.

American Gulf Pipefish
Syngnathus scovelli

Siphostoma scovelli Evermann & Kendall, 1896. Shamrock Point, Corpus Christi, Texas.

West Atlantic, Florida, Gulf of Mexico and south to Brazil. 16 trunk rings. Snout 2.3 in HEAD LENGTH, depth 3.7 in length. Length to 185 mm.

Incubation 11–15 days. 60–100 eggs. Coastal waters and mainly river drainages. Only member in genus that can breed in freshwater.

S. scovelli. Aquarium. Scott MICHAEL.

Keeled Pipefish *Syngnathus carinatus*

Siphostoma carinatum Gilbert, 1892. Northern Gulf of California.

East Pacific, endemic to type-locality. Length to 23 cm.

S. carinatus. Baja California, México, Brooding male, 20 cm below. After DAWSON.

Chocolate Pipefish *Syngnathus euchrous*

Syngnathus euchrous Fritzsche, 1980. Bahia Todos Santos, Baja California, México.

Eastern Pacific from the California region. May comprise two species. Lives in eelgrass beds and loose algae. Length to 30 cm.

S. euchrous. La Jolla, California. Female, length 15 cm. After DAWSON.

Barcheek Pipefish *Syngnathus exilis*

Siphostoma exile Osburn & Nichols, 1916. West San Benito I., Baja California, México.

Eastern Pacific, California region. In loose weeds. Length to 17 cm.

S. exilis. Baja California, México, Female, 15 cm below. After DAWSON.

ADDITIONAL SPECIES OF *SYNGNATHUS*
ILLUSTRATIONS NEEDED

Schmidt's Pipefish *Syngnathus schmidti*

Syngnathus schmidti Popov, 1927. Black Sea.

Subtemperate.

Caribbean Pipefish *Syngnathus caribbaeus*

Syngnathus caribbaeus Dawson, 1979. Fox Bay, Colón, Atlantic Panama.

West Atlantic. Similar to pelagicus/floridae. Length to 23 cm.Antilles to Venezuela. Muddy substrates near reef and *Thalassia* beds. Intertidal marine to 6 m.

Florida's Pipefish *Syngnathus floridae*

Siphostoma floridae Jordan & Gilbert, 1882. Pensacola Bay, Florida.

West Atlantic. Florida to Panama, also Bermuda, the Bahamas. BROAD PREORBITAL. Snout 7 in length. Lines along posterior tail rings. Inshore. Length to 27 cm. Brood with 800–1000 eggs.

Lego Pipefish *Syngnathus phlegon*

Syngnathus phlegon Risso, 1827. Nice, France.

Subtemperate Europe.

Dawson's Pipefish *Syngnathus dawsoni*

Micrognathus dawsoni Herald, 1969. Fish Bay, St. John. Virgin Islands.

West Atlantic, Greater and Lesser Antilles. Algae habitat, intertidal to about 7 m depth. Length to 17 cm. Similar to *S. pelagicus* but snout 8 in length and dark lateral body stripe.

Texas Pipefish *Syngnathus 'affinis'*

Syngnathus affinis Günther, 1870. Louisiana, U.S.A.

White to dark brown. Somewhat banded. Brown snout, white blotch below eye. Gulf of Mexico. Rapid decline of seagrass habitat has made this species very rare and caused possible extinction. Length to 22 cm.

Preoccupied by *Syngnathus affinis* Eichwald, 1831. Black Sea, Ukraine. =*S. abaster*?

Watermeyer's Pipefish *Syngnathus watermeyeri*

Syngnathus watermeyeri Smith, 1963.
Kasouga & Bushmans rivers, south-east Africa.

South Africa, only known type-locality region where in tidal zones of rivers. Inclusion to genus doubtful and provisional (Dawson, 1985). Differs in fewer pectoral fin rays and position of fin in relation to trunk ridges. The genus *Syngnathus* includes a number of doubtful genera that requires further studies, including the next species. Length to 13 cm.

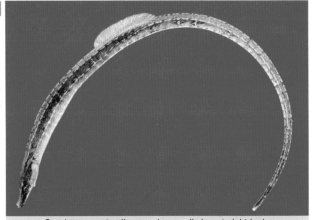

Specimen unnaturally curved, normally has straight body.

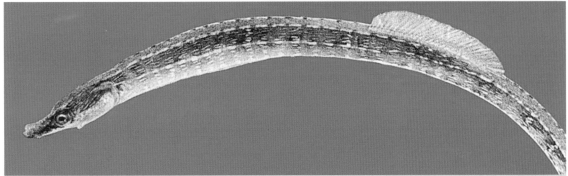

S. watermeyeri. Kleinemond River, South-eastern Africa. Phil HEEMSTRA.

Wreck Pipefish
Syngnathus safina

Syngnathus safina Paulus, 1992.
Gulf of Aqaba, Jordan.

Red Sea endemic. Lives at moderate depths, about 20–30 m, and inhabits low reef and rubble habitats or adjacent sand flats. Apparently a small species, known from few specimens less than 10 cm long.

Provisionally placed in genus by Paulus.

A

S. safina. Gulf of Aqaba, Jordan, Red Sea. Thomas PAULUS.

B

S. safina. Gulf of Aqaba, Jordan, Red Sea. Thomas PAULUS.

197

Masculine. Type species: *Bryx veleronis* Herald, 1940. Small genus with two Atlantic, one east Pacific, and one species in the West Indian Ocean and Red Sea. Distinguished from *Syngnathus* by the lack of anal fin.

Photographs are needed. Contributions and additional information will be appreciated and credited.

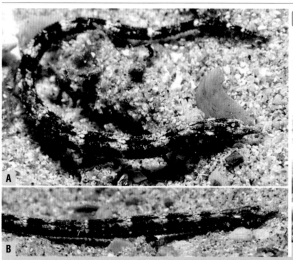

Florida's Pugnose Pipefish *Bryx veleronis*

Bryx veleronis Herald, 1940. Galapagos Islands.

Tropical to sub-tropical Eastern Atlantic. Sheltered rocky reefs with low algae growth. Small species, length to 60 mm.

B. veleronis. Panama, East Pacific. Clay BRYCE.

B. veleronis. Adult male, length 51 mm. After DAWSON.

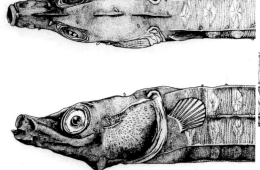

B. dunckeri. Male, length 8 cm. After DAWSON.

American Pugnose Pipefish
Bryx dunckeri

Syngnathus dunckeri Metzelaar, 1919. Dutch West Indies.

Argentina. Algae and rock bottom to 30 m. Also in estuaries. Length to 10 cm.

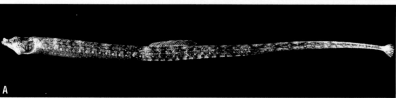

B. dunckeri. Male, length 8 cm. Various aspects of the head and trunk to tail section. After DAWSON.

Macran Pipefish *Bryx analicarens*

Syngnathus (Parasyngnathus) analicarens
Duncker, 1915.
Macran coast, Baluchistan.

West Indian Ocean and Red Sea. Rocky tide pools
and algal reefs to 45 m. Length to 135 mm.

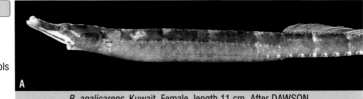

B. analicarens. Kuwait. Female, length 11 cm. After DAWSON.

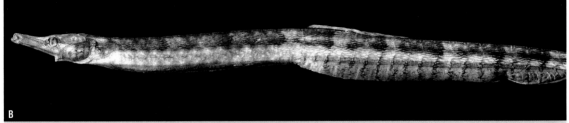

B. analicarens. Syntype. Male, length 128 mm. After DAWSON.

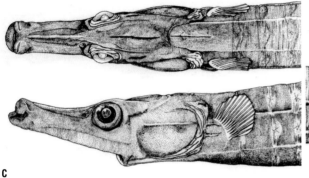

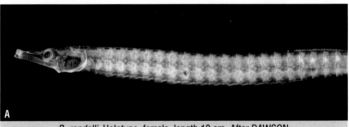

C

B. analicarens. Kuwait. Female, length 11 cm. Various aspects of the head and trunk to tail section. After DAWSON.

Ocellated Pipefish *Bryx randalli*

Syngnathus randalli Herald, 1965.
Isla Venados, Venezuela.

West Atlantic, Haiti, Lesser Antilles, Belize,
Provindencia Islands and eastern Venezuela to Golfo
de Cariaco. Shallow sub-tidal to 30 m depth. Rocky
and coral or mixed invertebrate substrates. Brood
contains about 60 eggs. Length to 10 cm.

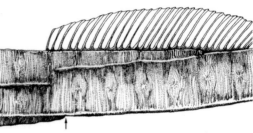

B. randalli. Holotype, female, length 10 cm. After DAWSON.

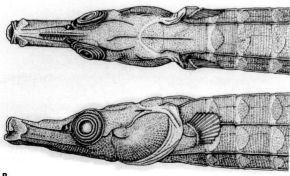

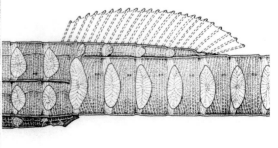

B

B. randalli. Holotype, female, length 10 cm. Various aspects of the head and trunk to tail section. After DAWSON.

Feminine. Type species: *Syngnathus argus* Richardson, 1840. An Australian and New Zealand endemic with at least 4 species, one of which only found in New Zealand, and two are unique to southern Australian waters. *Syngnathus*-like, but tail slender and used to wrap large posterior section around seagrasses, not in short coils such as the prehensile tail of seahorses. Dorsal fin large. Males brood pouch below anterior part of tail.

Spotted Pipefish
Stigmatopora argus

Syngnathus argus Richardson, 1840. Uncertain locality, probably Tasmania.

Widespread along Australia's south coast from central New South Wales to Shark Bay, WA. Occurs in small groups in *Zostera* beds in sheltered habitats, mainly estuarine, but may get washed out to sea in floating weeds. Length to 28 cm.

A *S. argus.* Brooding male, young visible through skin of pouch. Sorrento, Port Phillip Bay, Victoria, Australia.

B *S. argus.* Flinders, Western Port, Victoria, Australia.

C D *S. argus.* Females. Flinders, Western Port, Victoria, Australia. Depth 3 m. Length ~24 cm.

Gulfs Pipefish
Stigmatopora olivacea

Stigmatophora olivacea
Castelnau, 1872. St. Vincent's Gulf,
South Australia.

Appears to be restricted to the southern Australian Gulfs. This species was
confused with *S. argus* because of
the similar female phase. Snout much
more compressed and deep in males.
Locally common in seagrass beds.
Length to 22 cm.

A

S. olivacea. Close-up of male. Cape Jervis, South Australia.

B

S. olivacea. Male, about 20 cm long. Cape Jervis, South Australia.

C

S. olivacea. Female green and similarly spotted as *S. argus,* with brownish male. Port Victoria, St. Vincent Gulf, South Australia.

D

S. olivacea. Female. Port Victoria, St. Vincent Gulf, South Australia.

Wide-bodied Pipefish *Stigmatopora nigra*
Stigmatophora nigra Kaup, 1856 Tasmania.

Widespread along Australia's south coast from southern Queensland to Shark Bay, WA, and New Zealand. Male very slender, females develop a very broad trunk and display bright red stripes ventrally during courtship. One of the most abundant species in coastal seagrass beds. Congregates in small to large groups in *Zostera* beds and algal-reef habitats in sheltered bays and estuaries. Sometimes with loose weeds on open substrates. Length to 28 cm.

S. nigra. Aggregation, Seal Rocks, New South Wales, Australia.

S. nigra. Dunwich, Moreton Bay, Queensland, Australia. Bruce COWELL.

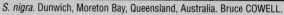

S. nigra. **C** Female above, displaying to male below. **D** About to spawn, pouch of male is pale expanded area below anterior tail section. Note broadly widened trunk of female. Portsea, Port Phillip Bay, Victoria, Australia.

S. nigra. Displaying female. Portsea, Port Phillip Bay, Victoria, Australia.

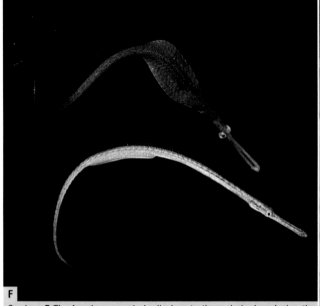

S. nigra. **F** The female aggressively displays to the male by broadening the trunk, swimming in circles around the male, as shown above, than suddenly shows a bright ventral side **G**. The male responds by opening the pouch and they embrace to spawn (**D**). Like with most syngnathids, the female is the one that initiates courtship and the more colourful sex. **G**

New Zealand Smooth Pipefish
Stigmatopora macropterygia

Syngnathus macropterygia
Duméril, 1870.
Oceania (=New Zealand).

New Zealand endemic, mainly known from the east coast of the southern Island, southern tip of northern Island, and south to Auckland Islands. Found in rockpools to 10 m depth in algal habitats. Males plain golden to grey or dark brown. Females with spots on top of trunk and blue dashes on sides. Length to 37 cm.

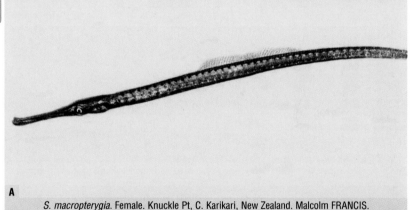

A

S. macropterygia. Female. Knuckle Pt, C. Karikari, New Zealand. Malcolm FRANCIS.

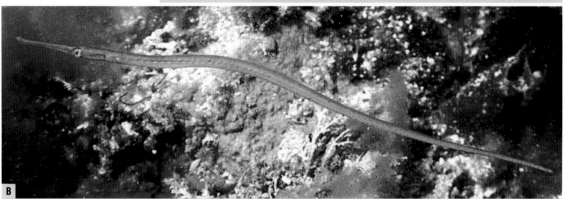

B

S. macropterygia. Probably male. Deep Water Cove, New Zealand. Malcolm FRANCIS.

Masculine. Type species: *Ichthyocampus davaoensis* Herald, 1953. Indo-Pacific genus with 3 known tiny species. Small and secretive in corals, apparently in association with *Galaxea* spp, and more species can be expected. Long pelagic juvenile stage and species probably widespread.

A

B. brauni. North West Cape, WA, Australia. Female, 54 mm (holotype). From dendrophylid coral at 10 m. Jerry ALLEN.

Braun's Pughead Pipefish *Bulbonaricus brauni*

Enchelyocampus brauni Dawson & Allen, 1978. NW Cape, WA, Australia.

Eastern Indian Ocean, Palau and Iriomote I., southern Japan. In association with corals to about 10 m depth, reported in Japan to over 20 m. Known to associate with *Galaxea musicalis* (Pipe-organ Coral) and *G. fascicularis*. Its white snout resembles the coral polyp and it may actually feed on them. Length to 72 mm.

B

C

B. brauni. Iriomote I., Ryukyu Is., Japan. Length ~65–70 mm. In association with *Galaxea fascicularis*. Korechika YANO.

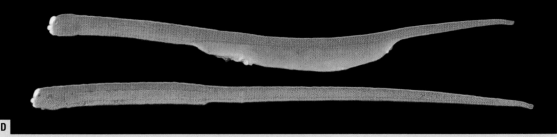

D

B. brauni. Iriomote I., Ryukyu Is., Japan. Male, 64.3 mm & female, 72.1 mm. From 15 m depth. Toshiyuki SUZUKI.

B. davaoensis. Male, 42 mm. After DAWSON.

Davao Pughead Pipefish *Bulbonaricus davaoensis*

Ichthyocampus (Bulbonaricus) davaoensis Herald, 1953.
Gulf of Davao, Mindoro I, Philippines.

Appears to be widespread West Pacific and doubtfully the same from eastern Africa where some specimens were collected from the coral *Galaxea fascicularis*. Length to 43 mm.

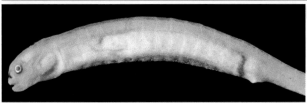

B. brucei. Paratype. Female, length 40 mm. After DAWSON.

Bruce's Pughead Pipefish *Bulbonaricus brucei*

Bulbonaricus brucei Dawson, 1984. Tanzania.

Only known from 6 specimens collected in the central lagoon at Pangani, Maziwi Is, Tanzania, where among the coral *Galaxea astreata* (reported as *G. clavus*) in 10 m depth. Length to 45 mm.

Only the family SOLENOSTOMIDAE is closely related to SYNGNATHIDAE. The next nearest groups are placed in the Sub-order AULOSTOMOIDEI. These include the families AULOSTOMIDAE, FISTULARIIDAE, CENTRISCIDAE, & MACRORAMPHOSIDAE. The related Order PEGASIFORMES, with the only family PEGASIDAE is included in full, and of the related GASTEROSTEIFORMES some members of the families AULORHYNCHIDAE & GASTEROSTEIDAE are included.

GHOSTPIPEFISHES - SOLENOSTOMIDAE

Represented by a single genus: *Solenostomus*. Differs from SYNGNATHIDAE with the presence of ventral fins, second dorsal fins, and all fins well developed and large. In contrast to SYNGNATHIDAE, the Ghostpipefishes have brooding females that use their large ventral fins to form a pouch. The fins are held together like flattened hands and the upper edge (or inner ray) of each fin hooks onto small ventro-lateral body spines.

GENUS *SOLENOSTOMUS* LACEPÈDE, 1803

Masculine. Type species: *Fistularia paradoxa* Linnaeus, 1770. Indo-West Pacific genus with about 6 species, some undescribed. Their similar morphology has led to much confusion in taxonomy. A revision of this group by Orr & Fritzsche, 1993, purely based on morphology and meristic values, failed to recognise several species and provided little progress. All species show a great degree of variability in colour and dermal appendages, as well as changes from postlarval to fully grown adults or between sexes. Most species appear to be widespread which is due to their long pelagic stage that reaches almost the full length of adults. The various species seem to have different preferences with regards to benthic habitats, ranging from muddy open substrates to rich invertebrate reefs. It appears that they are short-lived, perhaps having an annual cycle, as they appear seasonally in certain areas for the purpose of breeding. Adults generally pair up but some species maybe found in small groups. They are usually seen in sandy-rubble substrates near objects, floating almost motionless at near vertical angle to the bottom with the head down. They feed on small shrimp-like crustaceans that are either snapped up from the water column or from the bottom. Some species may target particular prey that live on other invertebrates such as on gorgonians or crinoids.

Aquarium kept specimens appeared to change sex as of the 5 individuals, 4 became brooding females. Two lots of these were collected as presumed pairs. This find was accidental and needs to be repeated to verify if males change to female when circumstances demand this.

Ornate Ghostpipefish *Solenostomus paradoxus*. Photographed off Sydney by Matthew BROOKE. This is the most commonly observed species in the genus. It is typified by the large number of spine-like dermal appendages that usually are numerous on the snout. This is a female with a brood in her pouch. Latter is formed by the ventral fins that are hooked up to small body spines.

A

S. paradoxus. Brooding female, about 11 cm long. Sydney Harbour, Australia.

Ornate Ghostpipefish
Solenostomus paradoxus

Fistularia paradoxa Pallas, 1770.
Ambon, Indonesia.

Widespread tropical Indo-West Pacific, from Red Sea and all of Indian Ocean to West and Central Pacific, and ranges into sub-tropical zones. Mostly pelagic until settling on substrate for breeding. Post-pelagic are almost fully transparent and more slender compared to those established to benthic life. Variable in colour from black to red and yellow, usually in a mix of bands and spots as shown in photographs. Usually settles along reef edges in current-prone areas where they pair or form small groups of several individuals. Up to 6 have been seen together with black-coral bushes on rubble ridges. In tropical waters they are often seen near crinoids or gorgonian corals that grow from reef over sand areas. They feed mostly on mysids but also target small benthic shrimps. Mostly occurs in sheltered coastal waters and estuaries where in channels to about 35 m depth. Length to 11 cm

B

S. paradoxus. Probable male, about 8 cm. Tulamben, Bali, Indonesia.

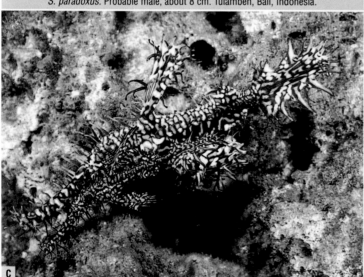

C

S. paradoxus. Small male, about 6 cm, in front of brooding female., Flores, Indonesia.

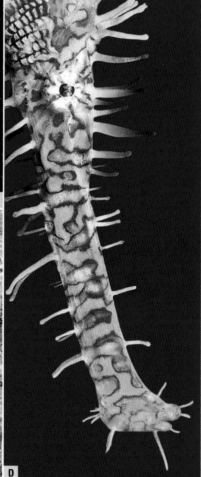

D

S. paradoxus. Maumere, Flores, Indonesia.

S. paradoxus. **E** Birth, young just expelled. **F** Refreshing eggs, opening and closing pouch. **G** Females, left individual in **E-F**. Flores, Indonesia.

S. paradoxus. Post pelagic stage, close-up. Bali, Indonesia.

S. paradoxus. Post pelagic stage, about 65 mm. Bali, Indonesia.

S. paradoxus. About 10 cm long. Milne Bay, Papua New Guinea.

A

S. cyanopterus. Maumere, Flores, Indonesia. Depth 10 m. Female, length 12 cm.

Solenostoma cyanopterus Bleeker, 1854.
Ceram, Indonesia.

Widespread tropical Indo-West Pacific, expatriating to sub-tropical zones. Mostly pelagic until settling on substrate for breeding. Primarily associates with vegetation, taking on colours of various algaes or seagrasses, from bright green to brown or black. Caudal peduncle short or lost in large females. Length to about 15 cm.

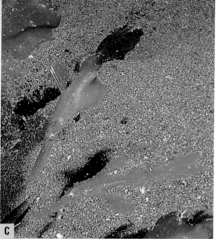

B

S. cyanopterus. Maumere, Flores, Indonesia. Depth 10 m. Male, length 12 cm.

C

S. cyanopterus. Bali, Indonesia. Takamasa TONOZUKA.

D

S. cyanopterus. Sydney, Australia. 15 cm.

E

F

S. cyanopterus. Bali, Indonesia. 14 cm. Brooding female, **F** close-up of pouch. Takamasa TONOZUKA.

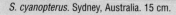

G. *S. cyanopterus.* Sydney, Australia. Female, 10 cm.

H. *S. cyanopterus.* Clovelly, Sydney, Australia. Length 10 cm.

I. *S. cyanopterus.* Manly, Sydney, Australia. Length 10 cm.

J. *S. cyanopterus.* Safaga, Red Sea. Depth 3 m.

K. *S. cyanopterus.* Post-pelagic stage. Seal Rocks, NSW, Australia. Length 8 cm.

L. *S. cyanopterus.* Milne Bay, PNG. Depth 15 m.

Delicate Ghostpipefish
Solenostomus leptosomus

Solenostomus leptosoma Tanaka, 1908.
Yodomi, Sagami Sea, Japan.

Widespread tropical Indo-West Pacific, known from Japan to Australia and also from Mauritius in the West Indian Ocean. Variable from pink to brown-red. Adult often with white streak along trunk. Usually an enlarged bushy fleshy appendage below middle of snout. Found along reef-edges, bordering on open sand-substrates, usually in depths of 15 m or more. Mostly pelagic until settling on substrate for breeding at almost full maximum size. Length to about 10 cm.

S. leptosomus. Bass Point, NSW, Australia. Depth 20 m. Length 8 cm. Post-pelagic stage.

S. leptosomus. Tulamben, Bali, Indonesia. Length 10 cm. Takamasa TONOZUKA.

S. leptosomus. Sydney, Australia. Length 9 cm.

S. leptosomus. Gravid female. Rodrigues, Mauritius. Frédérick DEPALLE.

Solenostomus paegnius
Jordan & Thompson, 1914. Misaki, Japan.

Widespread tropical Indo-West Pacific. Variable from bright-green to brown-red with small speckles, usually with several bushy fleshy appendages below snout. Similar to *S. cyanopterus*, but has short caudal peduncle. Algae-rubble reef and soft-bottom substrates, usually in depths of 10 m or more. Mostly pelagic until settling on substrate for breeding at almost full maximum size. Length to about 12 cm.

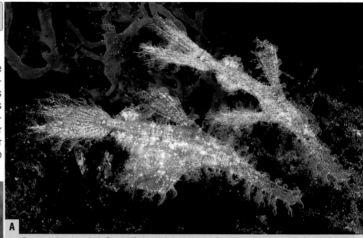

S. paegnius. Lembeh Strait, Sulawesi, Indonesia. D. 15 m. L. 11 cm. Roger STEENE.

S. paegnius. Sydney, Australia. 8-10 cm.

S. paegnius. Tulamben, Bali, Indonesia. Depth 15 m. Length 10 cm.

S. paegnius. **D** Oman. Phil WOODHEAD. **E** Sodwana Bay, South Africa. Greg DE VALLE.

Solenostomus armatus Weber, 1913.
Arafura Sea, Indonesia.

Only known from a few localities in the West Pacific. A rarely noticed species, plain greenish to yellow in colour and may look like the post-pelagic stage of the common *S. cyanopterus*. It differs most obviously by the long caudal peduncle and elongated caudal fin. Little is known about this species. The type material came from a muddy bottom trawl over 40 m depth. Length to 12 cm.

S. armatus. Pair, female with brood between pouched ventral fins. Osezaki, Japan. Masaya TAKAHASHI.

S. armatus. Pair, smaller and more slender male in foreground. Osezaki, Japan. Mie TAKEUCHI.

S. armatus. Close-up of pouch. Mie TAKEUCHI.

S. armatus. Female with brood, eggs clearly showing through the ventral-fin membranes. Length 12 cm. Osezaki, Japan. Mie TAKEUCHI.

Halimeda Ghostpipefish
Solenostomus sp 1

Widespread tropical Indo-West Pacific, known from numerous localities of the Indian Ocean to West Pacific. A small species with a very large head, its length up to about equal to length of the body. Caudal fin small, about same size and shape as the 1st dorsal and ventral fins. Caudal peduncle long, its length about 2–3 times its height. Often found with *Halimeda,* coralline or algae, matching their colours from bright green to red. Seen to about 15 m depth. Smallest species, length to 65 mm.

S. sp 1. Lembeh Strait, Sulawesi, Indonesia. Roger STEENE.

S. sp 1. Sulawesi, Roger STEENE.

S. sp 1. Tulamben, Bali, Indonesia. Depth 15 m. Length 6 cm.

S. sp 1? Milne Bay, PNG. Bob HALSTEAD.

S. sp 1. Great Barrier Reef, Australia. Depth 20 m. Length 65 mm. Phil WOODHEAD.

SHRIMPFISHES - CENTRISCIDAE

Small family, represented by two genera: *Aeoliscus* and *Centriscus* which can be visually distinguished by the structure of the large dorsal spine that is rigid in the first and hinged from a joint at about halfway in the second. They have adapted to a vertical posture and swim with their head downward. Body thin, wafer-like. They feed on tiny swimming invertebrates with their long pincer-like jaws. Eggs and young pelagic.

GENUS *AEOLISCUS* JORDAN & STARKS, 1902

Masculine. Type species: *Amphisile strigata* Günther, 1861. Indo-West Pacific genus with 2 tropical species, divided between Indian and Pacific Oceans. Commonly forms large synchronising schools over corals or seagrasses.

Spotted Shrimpfish
Aeoliscus punctulatus
Amphisile punctulata Bianconi, 1855-59. Mozambique.

West Indian Ocean and Red Sea. Recognised by the small black spots, sparsely distributed over the body. The dorsal-fin spine that is hinged (diagnostic for the genus) is clearly visible and represents the top when it is in vertical posture. Forms large schools over seagrass beds in sheltered zones, usually in just a few metres depth. Length to 15 cm.

A. punctulatus. Safaga, Red Sea, Egypt. Depth 5 m. Length 10 cm.

A. punctulatus. Safaga, Red Sea, Egypt. Depth 5 m. Length 10 cm.

Coral Shrimpfish
Aeoliscus strigatus

Amphisile strigata Günther, 1861.
Java, Indonesia.

Widespread West Pacific and Indian Ocean as far as the Seychelles. Recognised by hinged part of dorsal fin spine, typically as in **A**. Variable in colour with habitat. Greenish-yellow with diffused stripe when in seagrass environment. Pale with black stripe when found on open substrate with white sand or rubble. Usually found in large schools in coastal bays and estuaries over seagrass beds. Juveniles pelagic, settling at about 20 mm length and often amongst *Diadema*-urchin spines. Length to 14 cm.

A. *strigatus.* Derawan Kalimantan, Indonesia. Depth 3 m. Length 12 cm.

C. A. *strigatus.* Mabul, Borneo, Malaysia. Depth 7 m. Length 10 cm.

D. A. *strigatus.* Derawan Kalimantan, Indonesia. Depth 3 m. Length 12 cm.

E. A. *strigatus.* Maumere, Flores, Indonesia. Depth 1 m. Juvenile stage. Length 27 mm.

Masculine. Type species: *Centriscus scutatus* Linnaeus, 1758. Indo-West Pacific genus with 2 tropical species. Main dorsal spine without joint. Found in schools or small groups amongst corals or seawhip fields and seagrass beds.

Rigid Shrimpfish
Centriscus scutatus

Centriscus scutatus Linnaeus, 1758. East India.

Widespread West Pacific to Andaman Sea. Silvery with reddish brown to blackish mid-lateral stripe. Main dorsal spine moderately long and without joint. Usually in large schools amongst branching corals, seawhip gardens, and black coral bushes, to about 15 m depth. Small juveniles in surface waters and sometimes in small groups along beach edges in quiet bays and settle with crinoids or urchins. Length to 14 cm.

C. scutatus. Maumere, Flores, Indonesia. Depth 15 m. Length 8 cm.

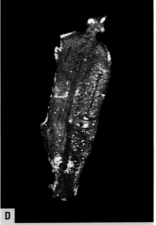

C. scutatus. Maumere, Flores, Indonesia. Depth 15 m. Length 8 cm.

C. scutatus. Flores, Indonesia. In urchin, at 10 m. Length ~6 mm.

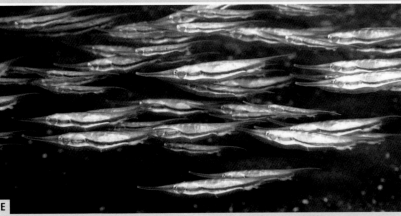

C. scutatus. Milne Bay, Papua New Guinea. Depth 10 m. Length 12 cm.

C. scutatus. Flores, Indonesia. Surface. Length ~5 & 9 mm.

Wafer Shrimpfish
Centriscus cristatus

Amphisile cristata De Vis 1885.
Noosa beach, Queensland, Australia.

Widespread West Pacific. Main dorsal spine rigid, without joint. Dusky to yellow mid-lateral stripe and adults with several blue bars along upper sides. Occurs solitary or in small groups in seagrass habitats, inshore or in estuaries. Largest shrimpfish, length to 20 cm.

A
C. cristatus. Flores, Indonesia. Depth 20 m. Length 4 cm.

B
C. cristatus. Derawan Kalimantan, Indonesia. Depth 3 m. Length 15 cm.

C
C. cristatus. Gilimanuk, Bali, Indonesia. Depth 5 m. Large adult, length ~20 cm.

Small family, represented by 3 genera, broadly distributed in temperate waters on continental shelves, two of which are found in the Southern Hemisphere only. Bellowsfishes, or snipefishes, have a rigid and highly compressed body, enlarged dorsal spine forming a spike, and a long tubular snout. Regarded as a subfamily of CENTRISCIDAE by some authors.

GENUS MACRORAMPHOSUS LACEPÈDE, 1803

Masculine. Type species: *Silurus cornutus* Forsskål, 1775. A single species with a temperate global distribution is presently recognised. It represents a complex of species and in some areas several are sympatric. Genus in need of revision.

A

Common Snipefish
Macroramphosus scolopax
Balistes scolopax Linnaeus, 1758.
Mediterranean Seas.

Widespread in all temperate seas, but comprises several similar species. Some populations are obviously different in body depth, and some occur sympatric. Variation within populations is moderate and lack of interest, being a trash fish, may have deterred studies until now. From the included photographs it is apparent that Australian and Japanese populations represent different species by their shape and colour. The true species lives in Europe in depths from 2–600 m where schools are sometimes observed by divers. In Australia the shallowest is about 60 m where two species are thought to occur. Length to 20 cm in Europe, 18 cm in Australia.

Remarks:
Slender form is often referred to as *M. gracilis*.

M. scolopax. Aquarium, Japan.

B

M. scolopax. Aquarium, Japan.

C

M. scolopax. Aquarium, New Zealand. Malcolm FRANCIS.

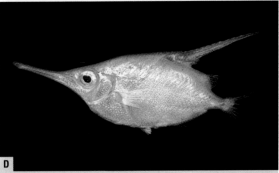

D

M. scolopax. Off Bermagui, NSW, Australia. Length 12 cm.

E

M. scolopax. Off Bermagui, NSW, Australia. Length 6 cm.

M. scolopax. The 'true' species, photographed in the eastern Atlantic by Peter WIRTZ.

Masculine. Type species: *Centriscus humerosus* Richardson, 1846. A single species, broadly distributed in the temperate waters of the southern hemisphere.

Banded Bellowsfish
Centriscops humerosus

Centriscus humerosus
Richardson, 1846.
Sea of South Australia.

Widespread Southern Hemisphere. Known from Australia, New Zealand, South America and South Africa. Juveniles silvery with faint bands that develop with growth. Adults have a large humped profile behind the head. The changes occur at near adult size and has led to confusion in recognising one or two species. A common bycatch in commercial fisheries. Lives near the bottom and recorded between 35 and 1000 m depth. Length to 28 cm.

A

C. humerosus. Off south-eastern Australia. Depth 500 m. Length 25 cm.

B

N. humerosus. Off south-eastern Australia. Depth 500 m. Length 18 cm.

C

N. humerosus. Off south-eastern Australia. Depth 500 m. Length 20 cm.

D

C. humerosus. South-eastern Australia. Depth 500 m. Length 10 cm.

E

C. humerosus. New Zealand. Malcolm FRANCIS.

Masculine. Type species: *Notopogon lilliei* Regan, 1914. Represented by 5 species, variously distributed in temperate and subtemperate southern seas. Benthic fishes, usually trawled in depths between about 100–1000 m.

Orange Bellowsfish *Notopogon xenosoma*

Notopogon xenosoma Regan, 1914.
Cape North, New Zealand.

Sub-temperate seas of the Southern Hemisphere, Australian southern mainland, northern New Zealand, South Africa, and southern Madagascar. Replaced by *N. fernandezianus* off the western and eastern coast of South America, which it was previously confused with. Adults identified by the orange colour and large dorsal spine that at least equals the length of the snout. Most specimens were captured between about 100 and 700 m depth. Length to about 18 cm.

N. xenosoma. Off south-eastern Australia. Depth 440 m. Length 18 cm.

Crested Bellowsfish *Notopogon lilliei*

Notopogon lilliei Regan, 1914. New Zealand.

Temperate seas of Australia, including Tasmanian region, and New Zealand. Adults identified by colour, low arched dorsal profile, and short dorsal spine that is much shorter than the length of the snout. Known depth range to about 600 m. Length to 27 cm.

J. MAY, CSIRO FISHERIES

A *N. lilliei*. Off Maria Island, Tasmania. Length 95 mm.

T. CARTER, CSIRO FISHERIES

B *N. lilliei*. Port Arthur, Tasmania, 54 m. Length 72 mm.

C *N. lilliei*. Off Maria Island, Tasmania. Length 22 cm. J. MAY, CSIRO FISHERIES.

D *N. lilliei*. New Zealand. Length ~25 cm. Malcolm FRANCIS.

Additional species of *Notopogon*

Notopogon armatus as *Centriscus armatus* Sauvage, 1879. Amsterdam & St. Paul I, southern Indian Ocean.
Notopogon macrosolen Barnard, 1925. Off Table Bay, South Africa.
Notopogon fernandezianus as *Centriscus fernandezianus* Delfin, 1899. Juan Fernandez, Argentina.

TRUMPETFISHES - AULOSTOMIDAE

Comprises a single genus. Body long and tubular, slightly compressed. The soft dorsal and anal fins are placed well back near caudal fin and ventral fins are placed about halfway along the body.

GENUS *AULOSTOMUS* LACEPÈDE, 1803

Masculine. Type species: *Fistularia chinensis* Linnaeus, 1766. Represented by 3 species, one in Pacific and two in Atlantic seas.

A

Pacific Trumpetfish *Aulostomus chinensis*
Fistularia chinensis Linnaeus, 1766. East Indies.

Widespread tropical Indo-Pacific, juveniles expatriating to sub-tropical zones. Readily identified by shape and colour. Maybe confused with flutemouth (Fistularidae) that is more slender and has a long filamentous caudal fin. A cunning predator of small fishes, swimming in the shade of harmless species (see **A**), matching their colour. Variable patterns shown. Various reef habitat, usually solitary. Length to 60 cm.

A. chinensis. Maldives. Matching colour of yellow goatfish, to sneak-up on prey.

B

C

A. chinensis. Maumere, Flores, Indonesia. Common adult colour.

A. chinensis. Flores, Indonesia. Large juvenile, ~30 cm.

South Atlantic Trumpetfish *Aulostomus strigosus*
Aulostomus strigosus Wheeler, 1955. St. Helena I., South Atlantic.

East Atlantic, Mauritania to Namibia and recently reported from Brazil. Reputed to change colour quickly when hunting. Mainly found in reef habitats, usually solitary but occasionally pairing. Length to 75 cm.

A. strigosus. Cape Verde Islands. Using grouper to hunt. Udo KEFRIG.

A

A. maculatus. Cayman, western Atlantic. Paul HUMANN.

West Atlantic Trumpetfish *Aulostomus maculatus*
Aulostoma maculatum Valenciennes, 1837. West Atlantic.

Tropical regions of the West Atlantic. Colour variations as the above species. Various reef habitat to about 25 m depth, usually solitary. Length reported to 90 cm.

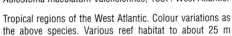

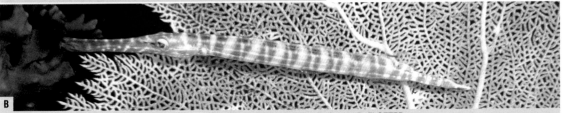

B

A. maculatus. Florida Keys, western Atlantic. Sergio R. FLOETER.

FLUTEMOUTHS - FISTULARIIDAE

Comprises a single genus. Body long and tubular, slightly compressed. The soft dorsal and anal fins are placed well back and ventral fins, with 6 rays, are placed about halfway along the body. Caudal fin forked with greatly prolonged central rays. Also called Cornetfishes.

GENUS *FISTULARIA* LINNAEUS, 1758

Feminine. Type species: *Fistularia tabacaria* Linnaeus, 1758. Represented by 4 tropical to sub-tropical species, 2 of which with global distribution, one in East Pacific and one in Atlantic.

Rough Flutemouth *Fistularia petimba*

Fistularia petimba Lacepède, 1803.
Equatorial Pacific.

Widespread Indo-West Pacific and Atlantic seas in tropical to temperate regions. Replaced in the Eastern Pacific by similar *F. corneta* Gilbert & Starks, 1904, Panama Market. Plain greenish brown over the back. Juveniles spotted. Broadly banded at night. Various reef habitats, juveniles entering estuaries. Usually seen singly or small groups. Adults usually in depths of 30 m or more. Sub-tropical, only found deep with cold upwellings in tropical regions. Identified by series of bony plates in front and behind dorsal fin. Length to 1.5 m.

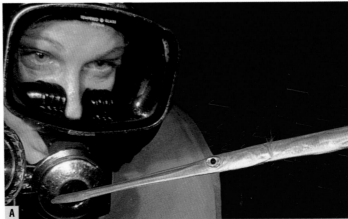

A

F. petimba. Adult observed by diver at night. Clovelly, NSW, Australia.

B *F. petimba.* Juv., night, Sydney, Australia. L. 35 cm.

C *F. petimba.* Juveniles, Seal Rocks, NSW, Australia. Length 25 cm.

D *F. petimba.* Juvenile. NSW, Australia. Length 20 cm. Spreading its two central caudal fin rays that has a large membrane in juveniles.

E

F

F. petimba. Adults, **E** daytime, **F** night time. Clovelly, NSW, Australia. Length ~ 1 m.

A

F. commersonii. Milne Bay, Papua New Guinea. Length 75 cm.

B

F. commersonii. Osezaki, Japan. Night. Length 65 cm.

C

F. commersonii. Maldives. Length 90 cm.

D

F. commersonii. Tulamben, Bali, Indonesia. Depth 6 m. Length 1 m.

E

F. commersonii. Tulamben, Bali, Indonesia. Depth 10 m. Length 75 cm. Night.

F. tabacaria. West Atlantic. Paul HUMANN.

Blue-spotted Cornetfish
Fistularia tabacaria

Fistularia tabacaria Linnaeus, 1758. America.

Widespread Atlantic seas. Found on reef flats and seagrass beds, usually to about 10 m depth. Series of blue spots on snout and body that become lines towards the tail. Generally shy and is difficult to approach, but sometimes curious and comes quite close when ignored. Length to 1.2 m, without the long filaments.

Order PEGASIFORMES

SEAMOTHS - PEGASIDAE

Represented by two genera, *Pegasus* Linnaeus, and *Eurypegasus* Bleeker. They feature highly depressed bodies and large horizontal pectoral fins that earned them their English name SEAMOTH. In addition, their body is protected by a structure of bony plates and rings, like in the Syngnathids. They have a rostrum that extends over and well in front of the mouth. Primarily benthic fishes, but have pelagic larval and early juvenile stages. Pairs rise from the substrate to spawn and produce pelagic eggs.

GENUS *PEGASUS* LINNAEUS, 1758

Masculine. Type species: *Pegasus volitans* Linnaeus, 1758. Indo-West Pacific genus with 3 species. *P. volitans* is widely distributed throughout the tropical region where juveniles maybe pelagic for a long time and carried far by currents. The other two have a limited distribution and live in slightly cooler regions. They take small invertebrates from the substrate that are sucked in whole by the protrusable mouth.

Slender Seamoth *Pegasus volitans*

Pegasus volitans Linnaeus, 1758. Ambon, Indonesia.

Widespread tropical Indo-West Pacific, with juveniles expatriating into subtropical regions. Usually in shallow depth to about 15 m, but reported to about 70 m. Adults mainly found in muddy estuaries where they pair. They may partly bury themselves in the sand or feed by slowly crawling around on their fins at various parts of the day. Sometimes found floating at surface. Easily recognised by long and slender shape. Colour variable grey, brown to black with spots or diffused broad bands on the tail. Length to 14 cm, rarely 18 cm (probably cooler regions only).

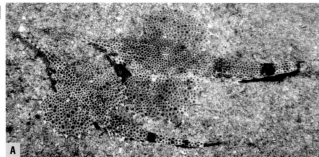
A · P. volitans. Pair, ~13 cm. Sulawesi, Indonesia. Depth 7 m.

B · P. volitans. Male, ~13 cm. Sulawesi, Indonesia. Depth 7 m.

C · P. volitans. Female, ~13 cm. Sulawesi, Indonesia. Depth 7 m.

D · P. volitans. Juvenile, ~5 cm. Sydney, Australia. Depth 5 m.

E · P. volitans. Pair, ~13 cm. Sulawesi, Indonesia. Depth 7 m.

F · P. volitans. Length ~10 cm. Queensland, Australia. Depth 3 m.

G · P. volitans. Juvenile, ~3 cm. Sydney, Australia. Depth 10 m.

Pegasus lancifer Kaup, 1861.
Type from Tasmania or Melbourne (Jawa given in error).

Endemic to sub-temperate Australian waters. Moderately common in sandy estuaries and congregates in certain areas to spawn. Often found along the sandy margins of seagrass beds where it often buries itself in the sediment. Courtship usually observed in spring and spawning peaks with big tides near dusk. Shallow protected zones to deep offshore, reported to 50 m depth. Length to 12 cm, usually 10 cm.

P. lancifer. **A** male, **B** female. ~10 cm. Melbourne, Australia.

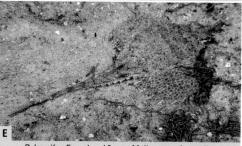

P. lancifer. Juvenile 25 mm. Melbourne, Australia.

P. lancifer. Male. ~10 cm. Melbourne, Australia.

P. lancifer. Female ~10 cm. Melbourne, Australia.

P. lancifer. Courting male. Melbourne, Australia.

P. lancifer. Courting male. Melbourne, Australia.

P. lancifer. Pectoral fin colour of courting male. Melbourne, Australia.

P. lancifer. Pair rises to spawn. Melbourne, Australia.

Brick Seamoth *Pegasus laternarius*

Pegasus laternarius Cuvier, 1816. Indian Ocean.

Mainland South-east Asian waters, ranging to southern Japan and in the Indian Ocean west to Sri Lanka. Rarely seen diving, except a few localities in Japan where they occur in sheltered muddy habitats. It has a variety of colours and can be dull to bright yellow or blue. Readily identified by its rather short rostrum when juvenile or female. It is more developed in males. Outside Japan they are mainly known from trawls between 30-100 m depth. Length to 8 cm.

A

B

E. laternarius. Juvenile. ~20 mm. Flores, Indonesia.

E. laternarius. Osezaki, Japan. Length 5 cm. Hiroyuki YAMAZAKI.

C

E. laternarius. Osezaki, Japan. Depth 23 m. Hiroyuki YAMAZAKI.

Little Dragonfish *Eurypegasus draconis*

Pegasus draconis Linnaeus, 1766. Ambon (Ex F. Ruysch, 1710).

Widespread Indo-West Pacific, expatriating to subtropical waters with pelagic young. Similar *E. papilio* (Gilbert, 1905) in Hawaii. Common short-bodied species in coastal waters. Adults usually in pairs on muddy substrates. Variable in colour, mimicking items such as shell bits in immediate vicinity for camouflage. Specimen kept in aquarium shed its outer layer of skin in a single piece to rid itself of epibiotic growth, such as algaes or invertebrates. Length to 85 mm.

G (next page) shows the mouth below the rostrum, and modified ventral fins adapted for walking.

Continues next page.

A

E. draconis. Close-up of **C** Female, 85 mm. Bali, Indonesia.

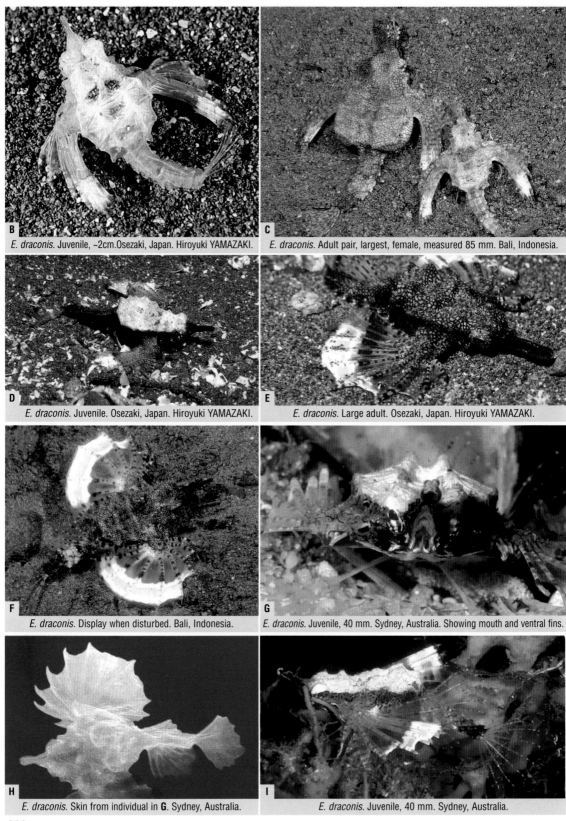

B   *E. draconis.* Juvenile, ~2cm.Osezaki, Japan. Hiroyuki YAMAZAKI.

C   *E. draconis.* Adult pair, largest, female, measured 85 mm. Bali, Indonesia.

D   *E. draconis.* Juvenile. Osezaki, Japan. Hiroyuki YAMAZAKI.

E   *E. draconis.* Large adult. Osezaki, Japan. Hiroyuki YAMAZAKI.

F   *E. draconis.* Display when disturbed. Bali, Indonesia.

G   *E. draconis.* Juvenile, 40 mm. Sydney, Australia. Showing mouth and ventral fins.

H   *E. draconis.* Skin from individual in **G**. Sydney, Australia.

I   *E. draconis.* Juvenile, 40 mm. Sydney, Australia.

Order GASTEROSTEIFORMES

Contains the families HYPOPTYCHIDAE, sand-lances, GASTEROSTEIDAE, sticklebacks, AULORHYNCHIDAE, tubesnouts, and INDOSTOMIDAE, lake-lances. Sometimes PEGASIDAE, the seamoths are included. As they have a close vicinity with the SYNGNATHIFORMES, selected representatives of the major two families are included for comparison.

STICKLEBACKS - GASTEROSTEIDAE

Represented by 5 genera of which *Gasterosteus* and *Spinachia* are well known, occurring in Marine and freshwater habitats and a species of each is included here. *Pungitius* has several species, and is similar to *Gasterosteus* but with many more dorsal fin spines and smaller, usually referred to as dwarf sticklebacks. North America genera *Apeltes* and *Culaea* (Whitley replaced the original name *Eucalia* as it was used in lepidoptera) appear to be monotypic including some variations.

GENUS *GASTEROSTEUS* LINNAEUS 1758

Masculine. Type species: *Gasterosteus aculeatus* Linnaeus, 1758. A widespread northern hemisphere genus, uncertain number of species with a single widespread form (included) and several variations that may be either specific or sub-specific. Fresh and brackish water.

Three-spined Stickleback
Gasterosteus aculeatus

Gasterosteus aculeatus Linnaeus, 1758. Europe.

Widespread coastal estuarine and freshwater regions of Europe, including British Islands and Scandinavia, and northern Pacific from Taiwan to the west coast of North America, as well as east Atlantic coast of Canada. Several sub-species. Locally common in various fresh and brackish water habitats, including small ponds or drains. Breeds in spring. Nuptial colour of male is blood-red on the belly and jaws, and green over the back. He builds a nest on the bottom and guards the eggs and young. Length variable with habitat and adults range from 5–10 cm.

Gasterosteus aculeatus. Aquarium, Amsterdam. Length ~7 cm.

GENUS *SPINACHIA* CUVIER 1816

Feminine. Type species: *Gasterosteus spinachia* Linnaeus, 1758. A marine genus, monotypic.

Fifteen-spined Stickleback
Spinachia spinachia

Gasterosteus spinachia Linnaeus, 1758. Europe.

Coastal marine regions of Europe, from Biscay in the south to central coast of Norway in the north. Shallow weed habitats, subtidal to about 10 m depth. Breeds in spring. Males usually larger than females and display with bluish colour to attract the female. He builds a nest of bits of weed and sticks in which he entices the female to spawn and deposit her 200 or so eggs in his nest. Length to 22 cm.

Remarks: This species shows some similarity with the aulostomids. Has become rare in some regions due to loss of seagrass habitat.

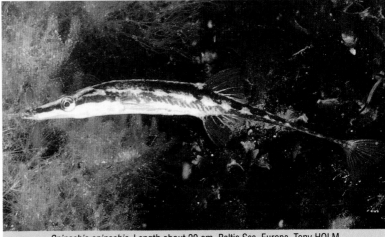

Spinachia spinachia. Length about 20 cm. Baltic Sea, Europe. Tony HOLM.

Represented by 3 monotypic genera, *Aulorhynchus*, *Auliscops* and *Aulichthys*. Latter included. Similar to sticklebacks, but much more elongated and numerous dorsal-spines.

GENUS *AULICHTHYS* BREVOORT 1862

Masculine. Type species: *Aulichthys japonicus* Brevoort, 1862. Monotypic, marine. Has about 25 spaced tiny dorsal fin spines.

Many-spined Tubesnout
Aulichthys japonicus

Aulichthys japonicus Brevoort, 1862
Japanese coast.

Northern Japan to the eastern Korean Peninsula. Shallow seagrass or weed covered reef habitats, forming large schools. Males are usually smaller than females, and blue-snouted in spawning season. They also have teeth on the jaws that are lacking in females. Length to 15 cm.

A *Aulichthys japonicus.* Yamaguchi Prefecture, Japan. Length 8 cm. Depth 5 m. Tomonori HIRATA.

B *Aulichthys japonicus.* Yamaguchi Prefecture, Japan. Length 8 cm. Depth 5 m. Tomonori HIRATA.

REFERENCES AND LITERATURE CONSULTED

STANDARD WORKS:
Allen, Gerald R. & Roger C. Steene, 1988. FISHES OF CHRISTMAS ISLAND INDIAN OCEAN
Allen, Gerald R. & Ross Robertson, 1994. FISHES of the TROPICAL EASTERN PACIFIC
Bleeker, P. 1849 to 1867. Various works, primarily on fishes of the Indonesian Archipelago, including contributions from various areas of the Indo-West Pacific, such as Japan and Australia. Has original descriptions of numerous syngnathids from marine and freshwater habitats
California Academy of Sciences, 1998. CATALOG OF FISHES. Volume 1–3
Dawson, C.E. 1976 to 1985. Numerous papers with original descriptions, generic revision and books on Syngnathidae from regions of the Indo-Pacific and Atlantic
Debelius, Helmut, 1997. MEDITERRANEAN AND ATLANTIC FISH GUIDE
Debelius, Helmut, 1998. RED SEA REEF GUIDE
Frickhinger, Karl Albert, 1991. FOSSILIEN-ATLAS Fische
Garrick-Maidment, Neil, 1997. Seahorses, conservation and Care. Kingdom Books England
Gloerfelt-Tarp, Thomas & Patricia J. Kailola. TRAWLED FISHES OF SOUTHERN INDONESIA AND NORTHWESTERN AUSTRALIA
Gomon, Martin F. et al, 1994. THE FISHES OF AUSTRALIA'S SOUTH COAST
Humann, Paul, 1997. FISCHFÜHRER KARIBIK
Kottelat, Maurice et al, 1993. FRESHWATER FISHES OF WESTERN INDONESIA AND SULAWESI
Kuiter Rudie H., 1992. TROPICAL REEF-FISHES OF THE WESTERN PACIFIC INDONESIA AND ADJACENT WATERS
Kuiter Rudie H., 1998. FISHES of the MALDIVES
Kuiter Rudie H., 1999. COASTAL FISHES OF SOUTH-EASTERN AUSTRALIA
Kuiter Rudie H., 1999. GUIDE TO SEA FISHES OF AUSTRALIA
Leis, J.M. & D.S. Rennis, 1983. THE LARVAE OF INDO-PACIFIC CORAL REEF FISHES
Leis, J.M. & T. Trnski, 1983. THE LARVAE OF INDO-PACIFIC SHOREFISHES
Lythgoe, John and Gillian, 1975. fishes of the sea
Lourie, Sara A, et all, 1999. SEAHORSES. Identification guide.
Masuda, Hajime, et al, 1984. The Fishes of the Japanese Archipelago.
Munro, I.S.R., 1956–1961. Handbook of Australian Fishes, instalments in *Fisheries Newsletter Sydney, N.S.W.*
Myers, Robert F., 1999. Micronesian Reef Fishes
Nakabo, Tetsuji, 1993. Fishes of Japan with Pictorial keys to The Species
Paulin, Chris et al, 1989. NEW ZEALAND FISH
Paxton et al, 1989. ZOOLOGICAL CATALOGUE OF AUSTRALIA 7 PISCES Petromyzontidae to Carangidae
Sainsbury et al, 1985. CONTINENTAL SHELF FISHES OF NORTHERN AND NORTH-WESTERN AUSTRALIA
Shao et al, 1992. Marine Fishes of the Ken-Ting National Park
Shen-Ed., 1993. Fishes of Taiwan
Smith, Margaret M. & Phillip C. Heemstra, 1986. Smith's Seafishes
Weber, M. and L.F. de Beaufort, 1913–1936. *The Fishes of the Indo-Australian Archipelago, 2–7*

SPECIFIC WORKS:
Duhamel, Guy, 1995. REVISION DES GENRES *CENTRISCOPS* ET *NOTOPOGON*, MACRORAMPHOSIDAE
Herold, Daphna, & Eugenie Clark, 1993. Monogamy, spawning and skin-shedding of the sea moth, *Eurypegasus draconis* (Pisces: Pegasidae)
Kuiter Rudie H., 2001. Revision of the Australian Seahorses of the Genus *Hippocampus* (Syngnathiformes: Syngnathidae) with Descriptions of Nine New Species. Records of the Australian Museum V53:293–340
Orr, James Wilder, & Ronald Fritzsche, 1993. Revision of the Ghost Pipefishes, Family Solenostomidae (Teleostei: Syngnathoidei)
Palsson, Wayne A. & Theodore W. Pietsch, 1989. REVISION OF THE ACANTHOPTERYGIAN FISH FAMILY PEGASIDAE (ORDER GASTEROSTEIFORMES)
Paulus, Thomas, 1993. Morphologie und Ökologie syntop lebender Syngnathidae (Pisces: Teleostei) des Roten Meeres
Pietsch, Theodore W., 1978. EVOLUTIONARY RELATIONSHIPS OF THE SEA MOTHS (TELEOSTEI: PEGASIDAE) WITH A CLASSIFICATION OF GASTEROSTEIFORM FAMILIES
Smith, J.L.B., 1963. FISHES OF THE FAMILY SYNGNATHIDAE from the Red Sea and the Western Indian Ocean. Ichthyological Bulletin No. 27. Rhodes University, Grahamstown, South Africa
Weber, M. 1913. *Die Fische der Siboga-Expedition*
Whitley, Gilbert, & Joyce Allan, 1958. THE SEA-HORSE AND ITS RELATIVES

234